# Spatial Temporal Information Systems

# Information Systems

An Ontological Approach Using STK®

# Spatial Temporal Information Systems

## An Ontological Approach Using STK®

Linda M. McNeil
T. S. Kelso

**CRC Press**
Taylor & Francis Group
Boca Raton London New York

CRC Press is an imprint of the
Taylor & Francis Group, an **informa** business

CRC Press
Taylor & Francis Group
6000 Broken Sound Parkway NW, Suite 300
Boca Raton, FL 33487-2742

First issued in paperback 2019

ISBN-13: 978-1-4665-0045-7 (hbk)
ISBN-13: 978-0-367-86701-0 (pbk)

---

**Library of Congress Cataloging-in-Publication Data**

---

McNeil, Linda M. (Linda Marie)
  Spatial temporal information systems : an ontological approach using STK / Linda M. McNeil, T.S. Kelso.
    pages cm
  Includes bibliographical references and index.
  ISBN 978-1-4665-0045-7 (hardback)
  1. Geographic information systems--Computer programs. 2. Spatial analysis (Statistics) I. Kelso, T. S. II. Title.

G70.212.M38 2014
910.285--dc23                                                                    2013031798

---

Visit the Taylor & Francis Web site at
http://www.taylorandfrancis.com

and the CRC Press Web site at
http://www.crcpress.com

# Contents

## Section II   STK Objects

# *Preface*

It wasn't too long ago that I remember sitting in the Mission Control Room at Wallops Island, Virginia, looking at the center screen as a rocket launch was in process. Minutes later, this same launch was again modeled within the STK software and shown by Jay Pittman, range commander and office chief at NASA Goddard Space Flight Center. Needless to say, I was blown away by the physics' dynamics and analysis. This was *rocket science*, visually stunning and with the capability of analyzing the ontology of the rocket and the objects around the rocket. It wasn't long after that time I began a new passion in my life: living and working in the modeling and simulation world of Spatial Temporal Information Systems using STK.

Later, through working on my master's degree in the world of geoscience using geographical information systems at Salisbury University, I realized just how powerful this software was. However, I was hard-pressed to find "how to" manuals that were readily available. Just before graduation, to my delight, my career took me directly to Analytical Graphics, Inc. where I became the DC Metro technical trainer. Here, my thoughts were confirmed: The clients in my classroom repeatedly told me how desperately they wanted a book to hold and guide them into the world of STK. It was with this thought that this book was born.

During this time, the geospatial community had been starving for the answers to strong 4D analytics for a decade and longer. The results of this bursting need gave birth to the Spatial Temporal Symposium that was presented at the American Association of Geographers in 2011, in Seattle, Washington. Here, in the plenary session, Drs. Douglas Richardson and Michael Goodchild laid the foundation for addressing this need. Spatial relationships have been studied in a variety of ways since ancient time. This study of ontology is part of the nature of the cartographic evaluations. Adding time allows us to understand what is happening with these relationships, the shapes of things as they morph. But how do we map that, should it be dynamic or attributal? In Richardson's presentation, he simplified it as the "Five Challenges of Spatial Temporal Analysis" (Space-Time Symposium, April 13, 2011):

- Spatial-Temporal Models
- Temporal Scale
- Ontology
- Real-Time/Real-World Interaction
- Analytical Tools

What is interesting when I heard these five challenges iterated was that I immediately understood how well STK handles all five of these challenges. It gave further credence to the notion that we needed to have more written material in the hands of the scientists so that people might understand Spatial Temporal Information Systems (STIS) and the framework of STK software. STK is not just for the rocket scientist; it is for the geoscientist, the astrophysicist, the engineer, the student, and anybody who has the need to answer physics-based, event-prediction questions. Although there are other forms of STIS used, none show the correlative understanding of ontological relationships as well as STK.

STK has been around since 1989; it primarily resided with the aerospace world of satellite application, so the geospatial community, as a whole, had very little knowledge of what STK can really do. In the last decade, this software has been able to not only maintain the de facto 4D analytics of Satellite Tool Kit but also has become the premier software in analyzing full multi-hop communications and other 4D types of ontology. This has made way for AGI to give the software a new name, Systems Tool Kit (STK).

This book is not intended to be a "how to" book regarding a particular software. AGI offers training courses to do that. At this time, these classes are free of charge and provide a wealth of information. This book, instead, is intended to extend the comprehensive training course. It is a study of the ontology of STK—which can easily be transferred to the study of other software systems to understand why they analyze the way they do.

There aren't a lot of algorithms in this book, deliberately. It is designed to be a high-level, approachable book for engineering college students as well as the PhD who needs further insight into STIS from an ontological perspective. It is expected that the reader has a background in physics or engineering to be able to fully understand some of the concepts; however, it can be used readily by the analyst sitting behind a desk who just needs more information on STK. In the future, there will undoubtedly be more books on the subject. These books will be deeper in concept and narrower in topic. However, for this first book, we now have a foundation to begin the study of STIS from an ontological perspective.

Knowing how well the software could meet these spatial temporal challenges has come through being a student of STK for many years. While I was teaching at Analytical Graphics, Inc., I used many phrases regarding STK. I think my favorite one was taken from the "As Seen on TV" commercials where "But wait, there is more!" truly applies to this software. STK is not just for rockets, satellites, or space. It handles communication, aircraft, ground vehicle, and ship modeling as well. Dynamically, it can handle all of these items together, all based over time or even in real time. This software is absolutely "video games for adults" in every form. Dr. Michael Scott, my mentor and good friend, once told me that "STK is the sexiest software around." I totally agree. I hope you do, too.

# *Acknowledgments*

I can't think of anybody who writes a book alone in a vacuum. We all need collaboration and a transference of knowledge. I would like to personally thank those people in my life who have made this book possible. This book is for you, your friends, and those who need a Spatial Temporal Information System like STK.

*Dr. T. S. Kelso:* As coauthor and collaborator for this book, T.S. has given insight and guidance for this book and much more. I have always enjoyed working with T.S. He is smart, funny, and most of all, he is a true astrophysicist. Thank you, T.S., for all you did in this book. Thank you for taking time from your busy life of conjunction analysis, conferences, and the never-ending saga of computer changes. Your work is amazing. There are times that I wish I had my PhD and could do what you do.

*Paul Graziani:* Paul is cofounder and CEO of Analytical Graphics, Inc. He also saw the need for this book and many more that are sitting on the back burner. Without his assistance, this book would not have been possible. AGI has been most gracious to me personally and to those the company employs. It honestly has been one of my favorite places to work. Thank you for the experience.

*Dr. Vince Coppola:* I love to learn. When you are around Vince, you are in a continual learning environment. One of my favorite things was to go in the back room on the third floor at AGI and dialog with Vince and Dr. James Woodburn. Vince spent a lot of time with me logically walking through how STK works. A lot of that information is distilled for you in this book. His insight into the algorithms, the functionality, and the physics helped me understand how to apply vast amounts of physics to Spatial Temporal Information Systems. His favorite application within the system is "interpolation." Vince collaborated with me on a white paper, "Spatial Temporal Analytics," written while I was still working at AGI. The white paper is the foundation of this book. Thank you, Vince. You're awesome.

*The other folks at AGI:* Joe Sheehan, Frank Linsalata, Todd Smith, Karen Haynes, Jonathan Lowe, Ed Gee, and the many more whose names I am not able to list. Thank you for all you have done. Rocket science just isn't as hard with your work.

Last, but most important, is an acknowledgment to my husband, *Warren McNeil.* Warren endured this book. The book took a lot of time away from our personal time with each other. When you reach empty nest time, enjoying alone time with your spouse is precious. Thank you for being gracious, kind, and supportive. I am so glad I married you.

# About the Authors

**Linda McNeil,** MSGIS-PA, is currently executive director of the Federation of Galaxy Explorers, a nonprofit space-based STEM educational program. Prior to this, she was a technical trainer for Analytical Graphics, Inc.; her primary function was to train professionals how to use STK software in multiple types of environments from DoD and Intel to NOAA, NASA, and more. She has a master's degree in geographical information systems and public administration from the University of Maryland's Salisbury  University. Linda has been working with GIS and other information systems for the past decade. She has 25 years of experience with computer science systems.

**Dr. T. S. Kelso** is a noted authority on satellite orbits. He is currently a senior research astrodynamicist for the Center for Space Standards & Innovation (CSSI), AGI's research organization that promotes public awareness of space information. He is also the webmaster of CelesTrak, a website dedicated to tracking space objects (including debris) and monitoring them for in-orbit collisions. A retired Air Force colonel with 31 years of active duty, Dr. Kelso served as the first director of the Air Force Space Command  Space Analysis Center (ASAC) at Peterson AFB in Colorado; led all Department of Defense analysis centers supporting the Columbia accident investigation; served as part of NASA's Near-Earth Object Science Definition Team; and consulted with the Massachusetts Institute of Technology to provide orbital models for the Hubble Space Telescope. During his career, he has held numerous teaching positions in the field of astrodynamics and has earned vast experience in research, analysis, acquisition, development, operations, and consulting. He is also an associate fellow of the American Institute of Aeronautics and Astronautics (AIAA).

# Section I

# The Basics

# 1

## The Basics of Spatial Temporal Information Systems

### Introduction

Spatial Temporal Information Systems (STIS) is a name (title) of computer systems with an emerging form of spatial analysis. An STIS is defined by the positions of objects within the environment, the use of dynamic time intervals, ontology or the study of the relationships of the objects, real-time or real world modeling, and lastly, the use of analytical tools. It is a blend of traditional Geographical Information Systems (GIS) with the use of Modeling and Simulation techniques. Our focus of this book is to reveal how an STIS works from the ontological perspective. Our approach is to show how an ontological relationship can be formed in an STIS by evaluating the objects and tools used within the environment. This is not a study of the algorithms used but a focus on how the objects and tools form relationships. This is a study of ontology as it is used within an STIS. The software used to create this study is Analytical Graphics' primary software, Systems Tool Kit® (STK).

An STIS is a system that includes spatial analytics but focuses on position and time. Just as ESRI's Arc software is a GIS, AGI's STK is an STIS. This book is about an STIS example using STK as we focus on how the objects and the tools work together to really understand the relationship of the position of objects over time. The use of ontology allows us to understand these relationships formed by the use of objects and tools. The focus of this book breaks down the ontology by discussing each component of the ontological relationship—the objects, the tools, and the output. This is where the value of the book will be to you, the reader. When you understand the theory of ontology as it is applied to the system, you can apply this to any spatial temporal system and understand spatial analytics with almost any system. The idea of using ontology is unique. Ontology is a database form of analysis. This approach changes the way many people look at a system.

Analytical Graphics, Inc. (AGI) makes Systems Tool Kit (STK)—a high-fidelity modeling and simulation (M&S) tool that allows analysts and engineers to model the spatial and temporal relationships between objects operating on the land, on or under the sea, in the air, and in space. STK provides an easy-to-use framework to define the properties of each object in this simulated environment and how it moves and is oriented over time. This framework allows the user to dynamically explore in depth the relationships among the objects.

As with any high-fidelity tool, understanding and mastering the tool can be a challenge. AGI provides an array of training to all STK users, but even a weeklong exposure only cracks the surface of the power of STK. College courses are being taught around the world to help the user understand the software and leverage the tools. From the industry user to the college student in the classroom, it is for such students that this book was written. The book is designed to help the dedicated STK user develop a deeper understanding of how STK works and the importance of the data being used within it to tackle everyday analysis tasks. This book extends the comprehensive training course that is taught by AGI. It explains more about how the software works from the computer science perspective of ontological relationships. This is a fundamental on-the-shelf reference guide. This book was written during the publication of STK Professional version 9.2.3 and glances into version 10. Although the book is written to represent STK in a universal way that will transcend versions and levels of software capabilities, all interpretations of object attributes and software capabilities are based on this version.

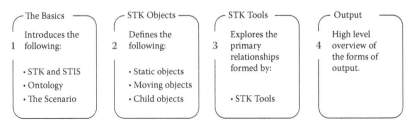

**Understanding STK using**
**Spatial Temporal Information Systems**

| The Basics | STK Objects | STK Tools | Output |
|---|---|---|---|
| 1 Introduces the following: | 2 Defines the following: | 3 Explores the primary relationships formed by: | 4 High level overview of the forms of output. |
| • STK and STIS<br>• Ontology<br>• The Scenario | • Static objects<br>• Moving objects<br>• Child objects | • STK Tools | |

**FIGURE 1.1**
Outline of book.

This book is divided into four main sections for easier understanding: The Basics, STK Objects, STK Tools, and Output. Part I, The Basics, is comprised of three chapters in the exploration of STK graphical user interface (GUI) navigation, identifying the basic parts within the software, and a guide on how to build a scenario. The STK Objects section, Part II, takes a detailed look at primary STK Objects. These chapters focus on how to define the object's position and other attributes. In addition, the STK Objects section takes a deeper look at defining an object's constraints and how these constraints affect analysis. The STK Tools section, Part III, gives insight to leveraging the computation of intervisibility, event detection, and signal evaluations. The final section, Part IV, Output, discusses graphics using static graphics, dynamic maps, reports, graphs, and the data providers that work with this form of output (see Figure 1.1).

## Understanding STK Basics

STK software has advanced analytical tools to help engineers and analysts understand line-of-sight event detection that occurs with objects both on Earth and in space. Aerospace Corporation's summary remarked that "access and visibility calculations were accurate to a high degree of confidence" (Aerospace Corporation, *Independent Verification and Validation*, 2000).

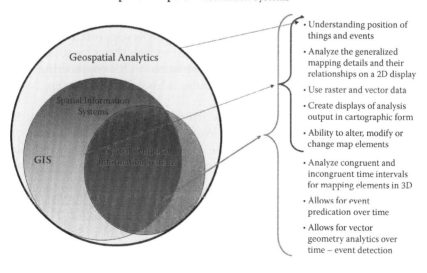

**Spatial Temporal Information Systems**

FIGURE 1.2
Spatial Temporal Information Systems.

STK is considered a Spatial Temporal Information System that models the position of objects at specific places and times. The dynamic interaction of objects combined with the event-detection tools allows the user to evaluate relationships from one or more objects to others over time. What sets STK apart from other modeling and simulation software are the time-dynamic event-detection capabilities. Cartographic output of these evaluations, as well as the ability to make movies and print graphs and reports, adds to the strength of STK. Formally, an STIS is defined as a specialized spatial information system that includes the element of time-based analysis. An STIS uses event prediction for objects, a 3D environment, and cartographic output (see Figure 1.2).

The STK interface provides an easy way to create objects and apply tools. Within STK, these objects and tools use physics-based modeling to answer questions and analyze specific time intervals within a variety of coordinate systems. Time intervals may be based on real or simulated time. Objects may be synchronized to the animation time clock defined within the software or have a user-defined time interval. Tools, which are used to evaluate the events or proximity of objects, use time intervals defined within an object or may be specifically user defined.

## The Workflows of STK for Ontological Studies

There are two different workflows used within STK: (1) the basic workflow of the software interface, and (2) the engineering workflow used to define the semantic level of the object, tool, and output attributes that allows for easy object and tool development. STK's basic workflow is supported by a graphical user interface (GUI) that guides the user through the steps to develop a scenario and allows for interactive manipulation. STK's engineering workflow allows the user to configure attributes, also called properties, for the objects and tools to enhance analysis.

The GUI is built modularly to allow user customization (see Figure 1.3). Because STK uses many of Microsoft Window's rich tools within the workspace, the user is able to configure the window positions and orientations to create a unique workspace environment. This allows the workflow to be customized also.

STK is analytical software that evaluates the relationships between real-world modeled objects and tools used to calculate line of sight, intervisibility, statistical variations, and signal analysis. The Engineering Workflow guides the user to define the STK Objects and STK Tools to the level of fidelity needed for each evaluation. When developing objects, the user uses the property arrangement to customize unique property attributes and make the object more realistic in physics capabilities and characteristics. Each

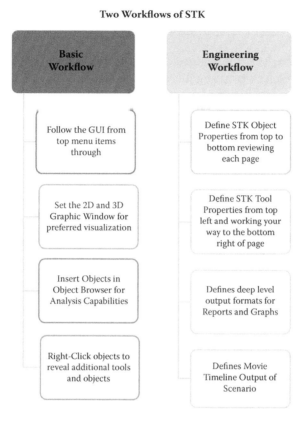

**FIGURE 1.3**
The basic workflow defines the environment of the scenario from a global perspective. The Engineering workflow is a systemic approach to defining the objects and tools in a more local manner using visual clues from the software features.

object has basic default parameters that allow computing for basic refinement. However, if more low-level, robust, and closer-to-real-life analysis is needed, then refining the properties and tools to match real-life characteristics is essential.

The STK object has a properties window with a list of several pages in the environment. Working these pages systematically from the top page down to the bottom is using the engineering workflow that is designed into the software. As these pages are modified to match a unique property, it allows the user to evaluate situations that simulate the real-life object it is modeling. Leveraging the engineering workflow allows users to easily develop ontological studies within the software. It is an approach to ease the use of how to develop scenarios, input objects, and use the tools within STK.

## Ontology

To have robust analysis, the engineering workflow is used to develop the ontology of the STK Objects by using the STK Tools. In the computer science world, ontology is the formal study of set domains and their attributes, as well as the relationships between these objects. In other words, it is the semantic-level evaluation of the relationships between objects, tools, and output as they are defined. STK allows you to create ontological studies in repeatable iterations or deeper refined versions to assist you in understanding real-world problems and the solutions STK shows you. As you review the objects and tools within the sections of this book, the ontology should become apparent (see Figure 1.4).

Reports and graphs are also refined using the engineering workflow and are the output of the ontological studies. Some of the ways tools vary are in analysis time intervals, use of light-time delay, or signal qualities. All of these items can be modified by using deep-level property changes found within the workflow of the tool or within the data providers from the Reporting and Graphs Manager. The engineering workflow will be further defined for both objects and tools during the course of this book.

STK is visually as well as analytically accurate as long as the STK Objects and the relationships with the STK Tools are defined with a high degree of fidelity. The STK Objects have attributes and constraints that refine the dynamics, kinematics, and capability. The STK Tools also may be refined by constraints,

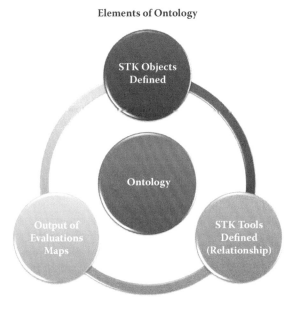

**FIGURE 1.4**
Elements of ontology.

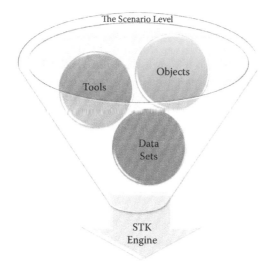

**FIGURE 1.5**
Dynamic STK.

step size, and methods of calculations. The attributes and constraints of both the STK Objects and Tools include data providers created to develop the computational data elements needed for the algorithms. In an integrated fashion, these elements of objects and tools and their defined properties result in an output evaluation. The time-dynamic geometry engine compiles the algorithms from these data providers of STK Objects and the relationships, as they are defined by the STK Tools. The output for these calculations is presented in reports, graphs, and visual responses (see Figure 1.5).

## Delineation between Objects and Tools

Within STK, the delineation between objects and tools is not overt within the GUI. Objects and tools are two different things by function. STK Objects are physical objects that have a position in three-dimensional space. STK Tools define relationships between the STK Objects that are usually based with the time-dynamic geometry engine or by tools that enhance the STK Object. Within versions 9.*n* and previous, STK has a section for STK Tools, Tools and Attached Tools (Children Tools), that is within the STK Object Browser and STK Tools that are a right-click off the objects. However, because objects and tools function differently, we have chosen to divide them in an obvious way in this book.

Objects are the object classes that represent items of the real world. These items may be a fixed point for a facility, city, town, or target. They may also represent moving vehicles, such as missiles, ground vehicles, ships, or satellites. Objects may be a region of interest as represented by the polygonal area target or may be a point representing central bodies, such as moons, planets, and stars. If there were a grammar structure within STK, we would call the object class a set of nouns.

Whereas, using STK grammar, the object class represents nouns, the tool class represents the verbs of the grammar. Tools classes are event-detection tools. Primarily, they analyze intervisibility, but they may also calculate proximity analysis and signal evaluations. Access is the primary tool that handles intervisibility calculations. It is the underlying calculation for most event-detection tools. Other tools are Deck Access, Chains, Coverage, Conjunction Analysis, Vector Geometry, Terrain Conversion, and forms of communications with signal processing. Because of the difference in functionality, tools have been separated to show how they are used and defined within STK at a very semantic level.

## Exploring the Objects

The STK Objects section provides valuable information about the robust nature of STK Objects. Objects within STK are used to model real-life buildings, equipment, or places within the software. The benefits of using the unique time-driven, object-orientated modeling within the STK environment become evident when attributes are defined on a semantic level for each object. The more realistic each object's attributes are in relation to the real-life object being modeled, the closer to real-life results ensue with the evaluations during the tool analysis, modeling, and simulation.

Objects are brought into the GUI environment using object-orientated methodology called encapsulation. Encapsulation is a class that allows the object to be accessed by an array of different methods depending on the level of access granted by the method. These different methods may or may not have unique attributes available. There are two main types of object methods that use encapsulation: the Scenario Object Selection Method and the STK Object Route or Position method (see Figure 1.6).

For instance, when you bring an aircraft into the STK Scenario environment, there is an array of Scenario Object Selection methods; one of them is "Insert Default." The default parameters within these properties are set for you to create a generic aircraft and define the properties at a very high level that only allows basic route waypoints to be selected. As you modify the STK Object Route properties, you may create an aircraft using Aircraft Mission Modeler (AMM). AMM is an enhanced modeler defined by aircraft type and

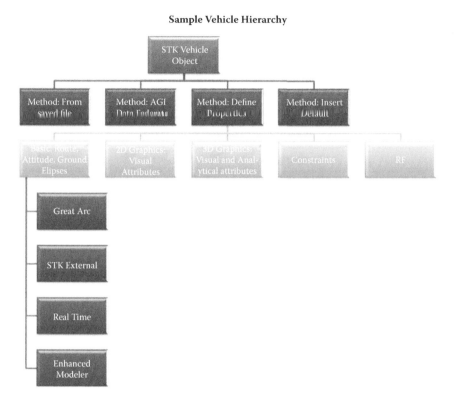

**Sample Vehicle Hierarchy**

**FIGURE 1.6**
Sample of the vehicle hierarchy.

mission. This level of method allows you to create a low-level model with a propagation method used to model real flight found within the route definition of the STK Aircraft properties.

There are two types of primary objects: parent and child. Through the use of object-orientated programming methods, inherency, the parent–child relationship is established. Inherency is an object class that allows a subtype object to use the rich attributes of the parent object and also include unique attributes of its own to model a specialized behavior. With STK, the parent objects, often called Scenario Objects, have a position, orientation, and time interval that is unique. Children, or Attached Objects, are a subtype to the parent object and by default utilize many of the characteristics of the parent object, such as position, orientation, and time interval (see Figure 1.7).

Object classes use encapsulation and inherency to allow STK Objects to be modified and made more realistic by changing the default parameters and matching the actual properties of an object more closely. The more refined an object is, the more accurately the analysis will match real-world scenarios. Object properties allow for deep-level analysis of events, such as intervisibility,

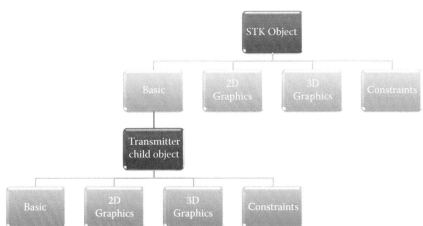

**FIGURE 1.7**
Parent–child hierarchy.

proximity, and qualitative evaluations. Propagators, or the predicted motion of an object, are calculated from within the properties of the object vehicle.

## Building Relationships with STK Tools

The STK Tools section evaluates the primary tools within STK and gives you a semantic-level understanding on how they work and what types of relationships they are used for. STK Tools build and evaluate relationships within the STK environment. This section explores what event-detection

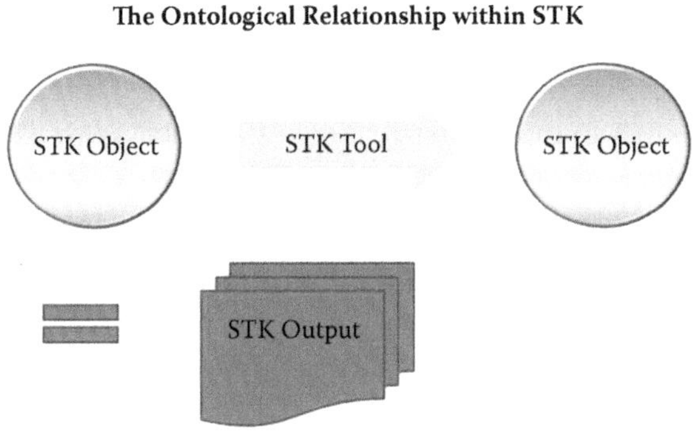

**The Ontological Relationship within STK**

**FIGURE 1.8**
Ontological relationships.

tools are and how they calculate object intervisibility, proximity evaluations, signal quality for single communications, and multihops. Tools within STK are specific and are determined by the events that are needed to be evaluated. Intervisibility events are used to calculate within tools such as Access, Deck Access, Chains, Communication Devices, and Coverage. Another event tool calculated is Conjunction Analysis. This is used primarily with orbiting space vehicles or objects. Most event tools use the Access interpolated algorithm as the basic form of computation (see Figure 1.8).

## Output

The last section of this book allows you to explore of the methods of output. Movies, static pictures, graphs, and reports are all forms of output within STK. As an intervisibility event is established, you can output reports and graphs that tell you the duration the objects can maintain the relationship, the time intervals in which they have the relationship, and even the look-angles used to establish the relationship. Cartographic output can be visualized in animation mode to simulate the object's movement over time or by creating static pictures of the events as they happen.

STK Tools evaluate the kinematics and dynamics of the STK Objects and how they interact with each other over a defined time interval. The interaction may be based on the proximity of the objects. It may consider the orientation of signal devices or the possibility of signal loss due to obstruction or degradation. Because space orientation and time intervals are critical for understanding, STK considers every object with its own coordinate system attached to the body frame to evaluate vehicle propagation and orientation evaluations and the application of constraints based on vector geometry of the object body. This tool capability gives relevance of time dynamics and physics applications to queries of, for example, when an object will be able to lock onto a signal or to understand how close objects are to one another. STK Tools evaluate the intervisibility, quality, and quantity of objects and signals. The output of this software allows you to visualize and analyze the objects and the tools from a modeling and simulation perspective.

## Data Providers

Data providers are the low-level attributes of the STK Object or STK Tool used to refine the output of an analysis. Components of the STK Object and STK Tool attributes are broken down into three primary types: Geometric, Time, and Calculations. They are hierarchical in format. Therefore, some of the property page attributes may contain time-interval information and vector-geometry-related information from the same page. Data providers are used heavily to create and customize data displays, reports, and graphs. They also are used to provide detailed verification and validation of the results. Data providers have the ability to break apart the computation algorithm and derive new algorithms at runtime of the STK Engine. Data providers are powerful. With version 10, expect to be able to leverage data providers better using the Time and Calculation Tools.

## What to Expect

As we begin to explore the world of STK, we can apply the approach of this study to many other software applications. Our goal is to understand the world of Spatial Temporal Information Systems. By using the approach of ontological studies, we can understand the implications of creating relationships between objects, defining the objects, and then modifying these objects (see Figure 1.9).

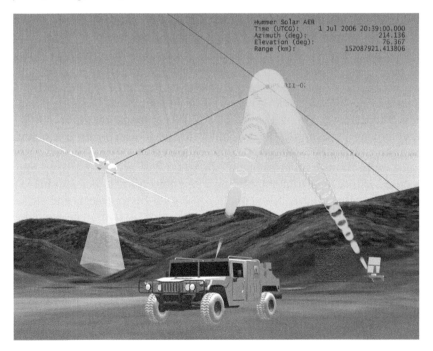

**FIGURE 1.9**
Example of object relationships.

# 2

## Ontology

**Objectives of This Chapter**

- Defining Ontology
- Understanding the Level of Properties
- Exploring STK Objects, STK Tools
- Exploring the 2D and 3D Object Windows

## Defining Ontology

Ontology in STK is used as a formalized study of the STK Objects, STK Tools, and the results from the analysis. The word *ontology* dates back to ancient Greek philosophy. The basic Platonic metaphysical meaning of the word is "the understanding and conceptualization of entities in categories and their generalizations." However, in the late 1900s, computer scientists such as Dr. Thomas Gruber from Stanford University modified this term as "the conceptualization analysis of objects and relationships in body of knowledge sharing and knowledge acquisition." In other words, ontology is the study of concepts or objects and their relationships within a domain. The use of ontology captures the data structures derived from the relationships and makes them visible. Dr. Gruber's work is currently governed by standards within the Resource Description Framework (RDF) and the OWL Web Ontology Language Guide for Computer Science. With STK, we use Dr. Gruber's definition of ontology to fit within the domain of the Spatial Temporal Information System (STIS) software.

Ontology within STK creates a focused level of study regarding how relationships are formed and defined between STK Objects. This study includes not only the semantic level of the STK Objects and the STK Tools that form the relationship but also the output of those relationships. The rigor of the study is completed by a full understanding of the semantic level of the STK Objects, including the methods and attributes of the object. STK Object attributes are defined by the properties and constraints. In addition to defining the STK Object, the STK Tool must also be defined. The tool defines the relationship between the STK Objects and builds the algorithms for a computed output or analysis result. The newly formed equations from the objects and the relationships are computed by the STK engine. The data, when computed, are visible to the user by the use of maps, graphs, reports, and simulation. When either the STK Object or the STK Tool has been modified, the data need to be recomputed and the output will reflect the changes from the results.

It is easiest if we think of each part of STK as individual sets of information within a domain (see Figure 2.1). For instance, the Root level of the software is the global domain of the software where the scenario level resides, whereas the STK Objects, STK Tool, and output (graphs and reports) are localized sets. The STK Objects are defined by attributes and constraints that modify their spatial relationship as they are analyzed and compared to other STK Objects. As mentioned before, the STK Tools are what define the set relationships among the STK Objects. These set relationships can be based on spatial position, distance, angles, orientation, and line of sight. After forming the relationships of the objects using the STK Tools, the primary algorithms are then computed using the STK Engine, allowing the output to be displayed. By creating set domains for each part of the software, we may refine the attributes of each entity within the sets and fully understand our analysis.

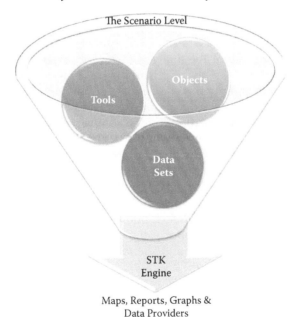

**FIGURE 2.1**
Dynamic STK.

Why does ontology really matter with STK? It matters because the more you understand the STK Object, the better your overall analysis will be. The computer science term for this is "GIGO." GIGO is simply stated as "Garbage In, Garbage Out." If you do not define your objects and tools to match the level of analysis you are looking for in modeling and simulation, then you will not get the quality of results you are looking for. The inverse is also true: If you take the time and define your object properties and tool properties, and correctly frame your output categories, then you are more likely to get the information you need. Sounds simple and it is. Many people assume computer software, no matter how well it is designed, will magically create the properties and give you the correct output with little or no effort by default. However, this is not true. We know that software is only a part of good analysis; user input is the other integral part. For example, if you want to recreate the orbit dynamics and performance of a spacecraft and have it match real life, you will need to select the correct propagation models and input the correct drag models, atmospherics and environmental effects, and object specifications. It requires some concentrated study and application—thus the need to apply ontology to the work environment of STK. Understand this allows you to understand the software better and gain quicker, more accurate results in your analysis.

To create an ontological study, you define both global and local properties within a scenario and then define the STK Objects and the STK Tools (see Figure 2.2). After computing and reviewing the output, you may find your analysis needs "tweaking." You might need to modify one or more of the

**The Approach to an Ontological Study**

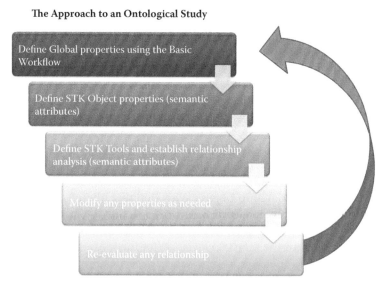

Define Global properties using the Basic Workflow

Define STK Object properties (semantic attributes)

Define STK Tools and establish relationship analysis (semantic attributes)

Modify any properties as needed

Re-evaluate any relationship

**FIGURE 2.2**
Defining ontology with STK.

objects or even the tool for more accuracy or to answer another question. With any change to either the object or the tools, the relationship and the output of the calculations might change. Therefore, the ontological studies within STK are reiterative in nature.

As objects change over time, the relationships may change. Default and user-defined properties influence how the intervisibility is defined. An object may go out of range and lose an established signal or the object may travel beyond the horizontal line of sight. An example would be the evaluation of a signal link between the ground station at Wallops Island, Virginia, and the International Space Station (ISS). Over the course of a day, Wallops would have multiple occasions during which it could track the ISS; this would define intervisibility. As Wallops tracks the ISS over the course of a day, it finds the time of intervisibility and provides an output of the information as you define it. The output may be in the form of a time interval and duration when the ground station is able to connect with the ISS, or perhaps it would come in the form of a graph or other visual output.

STK can predetermine the time intervals of intervisibility to make finding the connection time easier. The ability to find intervisibility uses an event tool called Access. The Access tool is the underlying event-detection tool used for most other tools within STK. For instance, signal evaluation depends not only on whether the objects have line of sight, or intervisibility, but also on whether the communication equipment can be seen and whether there are any limitations. As from the above example, the objects—both the ISS transmitter and the Wallops receiver dish—need to have Access before Signal can be evaluative for signal strength.

By using ontology as the first formal understanding of the relationship with the ISS, we begin to understand how the Wallops ground station as an object uses Access properties to build a relationship with the ISS. The ontology of these objects changed as constraints and signal communication equipment were added to these objects. To formally understand ontology and the software application, it is important to understand the software default parameters and how they may be modified and changed. To semantically understand the basic defaults and how to modify these defaults, STK interfaces give you visual clues within a structured engineering workflow to guide you in the discovery and development of your different localized object sets and parameters.

The ontological relationship begins with defining the global properties. When an STK instance is started, the user must define the of the instance, or scenario, the date and time intervals. This is the basic beginning to setting global properties from the scenario level. After these are set, the instance is initiated. This allows the user to refine the properties by editing the global preference settings and then adding any plug-ins to enhance the properties. The global settings become part of the overall equation on how the objects and tools interact within the scenario. (see Figure 2.3).

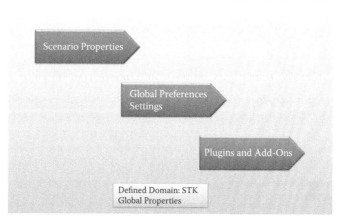

FIGURE 2.3
Scenario-level properties.

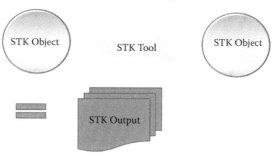

FIGURE 2.4
Creating a relationship.

Next, we input the objects and select the types of tools we need for analysis. We can test for high-level analysis to make sure we are using the correct objects and tools by computing this. There are times that the default parameters are considered good enough for the level of fidelity that is required for your evaluation (see Figure 2.4).

There are many times, however, that a deeper level of understanding is needed for a comprehensive review. This requires full disclosure of location, orientation, and performance of the objects and how they interact with well-defined tools. To do this, the objects and tools also need to be defined by each attribute to match the real-world processes. This becomes a formal ontological study when we begin to semantically define the attributes of both the objects and the tools to develop a robust form of output (see Figure 2.5).

The Ontological Relationship and the Object Properties Defined

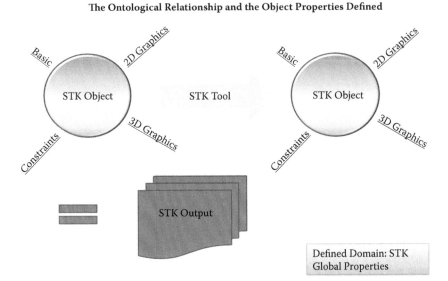

FIGURE 2.5
Defining attributes.

Finally, as a last step to the reiterative process of developing ontological studies, we can focus on the development of the output. We begin to analyze the output from these defined relationships, the user has the ability to tweak and refine the analysis by using the data filters built into the reports and graph section of the software. In this sense, we can use the an example of conjunction analysis. If we need to figure out the conjunction analysis of two objects in space, we are going to need to fully develop each object by defining the correct initial state of the object, including orientation. We also need to select the most effective propagator, define constraints of the objects, and then define the conjunction analysis tools to a level of refinement of distance analyzed, thresholds used for the analysis, and other data providers needed (see Figure 2.6).

## The Engineering Workflow

STK was built to improve the process of analysis and the quality of the results for spatial temporal analytics by using an engineering workflow within the software. In many agencies and working environments, multiple person-hours are wasted in redundant indigenous development of software, maintenance, and integration. Commercial off-the-shelf software, in general, is designed to eliminate the redundant programming work load, the software

The Ontological Relationship: Defined Objects and Tools with output

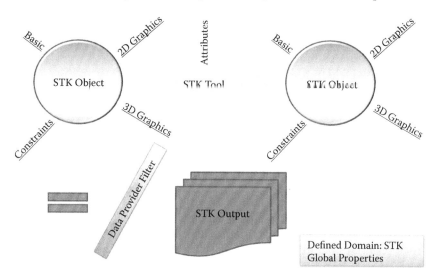

**FIGURE 2.6**
Defining all relationships.

testing, and the continual maintenance cycles used for in-house software development. STK was designed by engineers for engineers and analysts who need to produce more work and more analysis with less time using less overhead. The design of STK was created to also save time and money for the engineer. Rapid analysis development of scenarios within the graphical user interface (GUI) and the ability to customize the layout, click to visualize, and click to analyze are all part of the design (see Figure 2.7).

The Engineering Workflow of STK

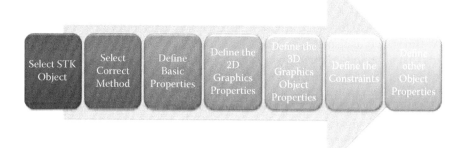

**FIGURE 2.7**
The engineering workflow of defining an object.

STK has rich, dynamic features that have been enhanced with the STK Help files. While most Help files within software applications are noninformative or weak in information, the STK technical writers have created a well-defined and extremely useful Help system. For most instances, if the user does not understand how STK defines its terms or what is on an attribute page, the Help button is usually available at the bottom of the page the user is viewing. In most cases, one click and the user is dynamically transported to the Help page they need to answer their questions.

## Understand the Properties

STK leverages the ontology of the modeling and simulation environment. STK is built by using attributes that set the default preferences of the software application in addition to three primary layers of objects with unique properties and defaults. The default preferences are found within the Edit/ Preferences option from the Edit task menu bar. In the preference menu, many of these preset defaults may be modified and set as new user-defined default values for the application. This is considered a global application change that becomes the new set of defaults. If changes are made to the preset values, the new user-defined defaults are maintained for subsequent times the application is opened.

Plug-ins and extensibility products for STK software such as Movie Timeline, Wind Farm, and Qualnet are all hooked into the software within the Edit/Preferences/UI Plug-ins page. MATLAB® users would define engine outputs within the preferences and propagator defaults that are set here. Another example of this would be to modify the Access defaults to not use Line of Sight as a default value for an analysis evaluating eclipse sessions. However, the analyst would need to be aware that this new default would be set for all other further STK sessions, even those that may have the need to have Line of Sight on for default calculations. To set and create your own plug-ins, use the AGI-provided document in Appendix A.

At times, the user needs to modify the path for object and terrain downloads that are handled through the software. The path could be locally modified during the download process every time a download is performed, or it may be permanently modified to a predetermined user-defined default within the Edit/Preferences/Find file page. Changing a global application default has benefits but can also cause issues. At this point, the user would need to decide if the value changes are needed to be global and permanent as defined within the Edit/Preferences section, or scenario-level global changes that only affect the global parts of each scenario as it is defined within the Root level. Many times, the changes needed are simply local changes that are made directly with the Object level or Tool level used for analysis.

Although global application changes make modifications to the software that become new default behaviors for the software every time the software is used, there are also other changes that can be made with the software that affect the properties of the individual scenario. These modifications are either global or local scenario changes that are handled within one of the four primary levels of set domains within STK.

## Four Primary Levels of Set Domains

The four primary levels of set domains within the software are the Root, STK Object, STK Tools, and Output levels. Each unique layer is able to refine its individualized properties to make the analysis more realistic and meaningful to the analyst or engineer. The Root-level properties allow for more global modifications to the entire scenario, whereas the Object, Tool, and Output levels are more localized. Tool-level attributes primarily affect the way the calculations are defined when evaluating the relationship between the objects. Reporting allows you to customize the reports, graphs, and mapping to clearly articulate the information of your analysis. Details on each level can be found within the Help files of the software (see Figure 2.8).

### The Root Level

This is the scenario level of the software from a global perspective that affects the current session or scenario. When the user creates a scenario, the scenario name creates the Root-level attributes within the software. Any calculations during the scenario time interval use the defaults and user-defined settings globally. There are four property sections of predefined default settings and

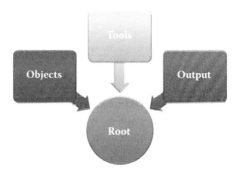

**Four Primary Levels of Set Domains**

**FIGURE 2.8**
Four primary levels of set domains.

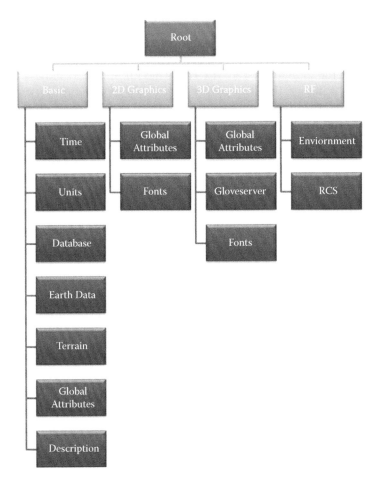

**FIGURE 2.9**
Root-level hierarchy.

attributes that may be modified within the Root: Basic, 2D Graphics, 3D Graphics, and RF (see Figure 2.9).

## Basic

Basic properties include the following modifiable pages:

   **Time:** *Scenario Time* intervals are set for both the Analysis and the Animation Time Period. This is the same information that was user defined with the *Start-Up Wizard*. The *Analysis Time Period* reflects the lifetime span interval of the analysis. This could last for just a few minutes, months, or even years. Calculation *time steps* are set to allow interpolated calculations to become manageable. For short 24-hour time intervals, having a small time step of every 60 seconds would seem valid and computation time effective;

 1 Jul 2007 00:00:00.000

**FIGURE 2.10**
Animation bar introducing Time.

however, it would be computationally expensive to maintain a 60 second time step for months or years of analysis. A common analysis time for monthly analysis is 24 hours of time step.

*Animation Time* is the time interval set for the animation clock. It will determine the time interval the user is able to visualize the objects within the scenario. The Animation Time is usually equal to or a subset of the Animation Time period. The time of your scenario should meet or exceed the time needed for each object's time interval. The *Epoch Time* is the time that your analysis will begin.

Defaults for the Analysis Time and Animation Time Period use today's date as captured from the computer's date/time clock. The interval of time for the default is set at 24 hours. The default time step is preset at every 60 seconds.

**Units:** The global default units for the scenario session are listed in Table 2.1 (all of which may be modified). If defaults need to be customized, modifications may be made. Getting to the information is tricky. The cell that the unit is modified at is dynamic. Only by placing the cursor and clicking on the unit's cell that requires modification will you be able to access the pull-down menu of items to make the necessary changes. These units may also be modified locally from the Object or Tool levels of the scenario. Table 2.1 represents the default values for the unit settings (all of which may be modified).

**Database:** The database page allows the user to update or modify the four primary database types—City, Facility, Satellite, or Star database files—or to add an auxiliary database type. Default databases are selected in the *Database Type* section, whereas, if these databases have been changed and have all the needed file sets, then they are inserted using the *Auxiliary* section. STK comes with the four default databases, but many times the user will have the need to customize these databases to make his/her own network. For instance, a NOAA contractor would need to know where all of his satellite and radar dishes were and use these facilities on a daily basis. It would make sense for the user to create a custom Facility database for this specific use.

STK databases are sets of files with specified file extensions that follow a predesigned STK format. Template information for these database files for each file type is currently found within the Help files under "Create & Import External Files." Most, but not all, custom data files have file version numbers and keywords used within the grouping. The version numbers of the files correlate with the version number of the software release. The keywords groups are standard reserved words—primarily, BEGIN <group name> and END <group name> file blocks. Comma delimitation is used

**TABLE 2.1**

Default Values for the Unit Settings

| Dimension | Current Unit |
|---|---|
| Distance | Kilometers |
| Time | Seconds |
| DateFormat | Gregorian UTC |
| Angle | Degrees |
| Mass | Kilograms |
| Power | dBW |
| Frequency | Gigahertz |
| SmallDistance | Meters |
| Latitude | Degrees |
| Longitude | Degrees |
| Duration | Hr:Min:Sec |
| Temperature | Kelvin |
| SmallTime | Seconds |
| Ratio | Decibel |
| RCS | Decibel (dBsm) |
| DopplerVelocity | Meters Per Sec |
| SARTimeResProd | Meter-Second |
| Force | Newtons |
| Pressure | Pascals |
| SpecificImpulse | Seconds |
| PRF | Kilohertz |
| Bandwidth | Megahertz |
| Small Velocity | Centimeters per Second |
| Percent | Percentage |
| MissionModelerDistance | Nautical Miles |
| MissionModelerAltitude | Feet |
| MissionModelerFuelQuantity | Pounds |
| MissinModelerRunwayLength | Kilofeet |
| MissionModelerBearingAngle | Degrees |
| MissionModelerAngleOfAttack | Degrees |
| MissionModelerAltitudeAngle | Degrees |
| MissionModelerG | Standard Sea Level G(G-Sea Level) |
| Solid Angle | Steradians (Sterad) |
| MissionModelerTSFC | TSFC LbmHrLbf |
| MissionModelerPSFC | PSFC LbmHrHp |
| MissionModelerForce | Pounds |
| MissionModelerPower | Horsepower |
| SpectraBandwidth | Hertz |
| Bits | MegaBits |
| RadiationDose | Rads |
| MagneticField | nanoTesla |
| RatiationShieldThickness | Mils |
| ParticleEnergy | MeV |

for values; however, the unit labels are usually predefined within the Root Level Units page. For version 9, the default database files are found at C:\ProgramData\AGI\STK 9\Databases\. For each of these main databases, the rows are always considered as one object and the columns always begin in the zero (0) column.

**Earth Data:** The default Earth Orientation Parameters (EOP) file is specified here and may be modified if needed. The default file is provided by CelesTrak and uses International Earth Rotation Service (IERS) and U.S. Naval Observatory (USNO) data. This file, as well as Space Weather data, are critical to the calculations used for satellite propagation, vehicle motion, and aerodynamics. Deltas in the source information are small; however, if the user needs to calculate with a user-defined data source, this may be modified in the Filename browsing box by selecting the "Ellipsis" button, then click the "Reload EOP File" button. For more information on EOP data, please refer to CelesTrak.com and search for a paper entitled "Using EOP and Space Weather Data for Satellite Operation," written by David A. Vallado and Dr. T. S. Kelso.

**Terrain:** Adding Terrain files in STK allows for better visualization and flight dynamics, the creation of azimuth-elevation masks, and the ability to look for line-of-sight impedances caused by terrain. The Terrain page allows terrain data files to be uploaded in this area. These terrain files are organized by central bodies, such as Earth or Mars. STK uses Terrain visually and analytically. Both are handled differently within the software. For analytical purposes, standard National Geospatial-Intelligence Agency (NGA) and U.S. Geological Survey (USGS) file types may be directly read when hooked into the software. However, to read the terrain files visually, the software must be converted to an indigenous file type created by AGI for terrain visualization that uses a *.pdtt extension. The training on converting terrain files is covered in the comprehensive STK courses with AGI. Further discussion on the differences between analytical and visual terrain files and how to convert files for visualization can be found in the methodology section of this book.

Sources for Terrain vary from department to department. Most companies and agencies already have a large terrain file repository established. However, if a person needs to collect terrain information for specific projects, the best sources for terrain files are:

1. National Geospatial-Intelligence Agency (NGA)—www.nga.mil

   https://www1.nga.mil/ProductsServices/Pages/PublicProducts.aspx

2. United States Geological Survey (USGS)—www.usgs.gov

   http://www.usgs.gov/pubprod/maps.html

3. Analytical Graphics—www.agi.com

   http://www.agi.com/products/by-product-type/applications/stk/add-on-modules/stk-terrain-imagery-maps/

There are other places to obtain terrain data; however, the above sources are typically vetted sources. There is individual support for other terrain types such as Polar Stereographic Projections formats (PDS). The user will need to contact AGI directly for this support.

**Global Attributes:** These properties give the user a warning message when using point mass vector objects such as missiles, satellites, and aircraft.

**Description:** This page is a reference page for metadata information. The long description is the same reference box used with scenarios and is started within the Welcome to STK start-up wizard. This may be modified at any time as the scenario evolves.

### 2D Graphics

2D Graphic properties include the Global Attributes and the Fonts pages. Both pages are modifiable. These pages primarily give the user a variety of choices on how he or she would like to view the objects and the texts within the 2D window environment. 2D Graphics enhance visualization. There are some specialized maps that include details for Earth as well as other the specific central bodies. Additional latitude and longitude lines may be shown and enhanced with a unique color schema for better analysis.

### 3D Graphics

3D Graphics properties handle both analytical and visual constraints. They include the following modifiable pages: Global Attributes, Globeserver, and Fonts. Global Attributes allows off-screen rendering for large graphic file types, if this setting has been allowed at the global applications level and if the computer hardware is capable of this task. Graphics are layered in the 3D window with Terrain, graphics and line information as well as the STK Objects with the computed visual results of the STK Tool analysis (see Figure 2.11). Setting surface lines to draw directly on the terrain when the objects are created may be enabled here. Additionally, allocating computing cache amounts for imagery can be controlled here. Lastly, surface referencing of the Earth's globe is also maintained here. The default is to reference WGS84 ellipsoid and may be switched to using a reference of Mean Sea Level if needed. This setting can affect the smoothness of the aircraft or other point mass vector object as it is in flight. For some scenarios, this should be modified for optimum performance.

### RF

RF properties refer to the global scenario changes that may be made for radio frequency signal calculations. They include the pages for Environment and Radar Cross Section (RCS) options. In order for the software to consider introducing environmental change models to the signal calculations, these

**FIGURE 2.11**
Layer graphic.

models must be toggled on within the Environmental RF page. For most of these environmental models, STK has a chosen default model, and other models available for use with the added flexibility of being able to introduce custom in-house models by using a script plug-in. Environmental changes include using Rain Loss models to evaluate signal degradation. The primary versions used for Rain Loss are the ITU-R P618-9, which is the default, and the Crane 1985 model. Another option would be to use the Cloud and Fog Model. Specific Cloud and Fog models may also be used where the user defines the cloud ceiling, cloud layer thickness, cloud temperature, and cloud liquid water density. Tropospheric scintillation considerations may also be included. They could calculate for tropospheric fade outages from deep fade. For Atmospheric Absorption Models, STK uses Simple Satcom (default), ITU-R P676-5, and Two Ray models.

## Object Level

The Object Level within STK consists of local instances of objects with methods and classes that behave based on global parameters and local object parameter settings. In the STK environment, much like other spatial information systems, objects present themselves as vector or stationary data using geometric shapes such as points, lines, and polygons. You may analyze multiple instances of the same object within a scenario. The primary difference is the ability to include time as more than just a local attribute, instead using time as incongruent and congruent intervals that are integrated into the properties of the global and local aspects of the software.

Most objects are points, when strictly 2D is considered a simple vertex with an *x* and *y* coordinate. In 3D, there are two types of points: a fixed point and a point mass with a vector. When an object moves over time, the default is a visual polyline trailing after the point. Polygons are used to represent defined areas of interest, as in the use of countries, city boundaries, or area targets.

Introducing an object into the scenario is relatively simple. It may be approached several ways, but the most common approach is to use the "Insert STK Object" wizard (icon). When an object is introduced into the scenario, it will nest under the Root level of the object browser (see Figure 2.12).

There are two primary types of objects: parent objects and child objects. This parent–child relationship is considered hierarchical and allows for rules of inherency from the parent object to the child. A parent object may have many instances of children relationships, but an individual child may only have one parent from which inherency may be derived. Objects allow for intervisibility and ontological computations among other objects.

Child objects are also called Attached objects. These objects require a parent as a primary object. For instance, an antenna would require and inherit the position of facility, vehicle, or satellite. One primary inherited property from the parent is position. Child objects are primarily considered to be antennas, radars, receivers, sensors, and transmitters. The objects' properties, either parent or child, are specifically designed to meet high-level and low-level physics modeling based on the type of object that is being evaluated.

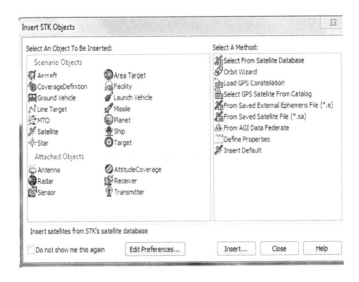

**FIGURE 2.12**
Objects.

Object property modifications are local to the object, but by the study of ontology we understand these object modifications can dynamically change many possible calculations as relationships are established. It is important for the user to understand the objects and their properties before relationships are established.

## Tool Level

Tools are the computational devices that allow objects to be evaluated. Currently, STK Tools are found in the Task Menu sections, the STK Objects wizard, and by right-clicking a specific object in the object browser. Tools are what answer the questions we are trying to ask with STK. Instances of tools come in many forms within STK. Most tools compute events, such as when objects have intervisibility, or they can define the quality or quantity found within evaluations. Tools are local properties, and changes within a tool only affect the ontological evaluations locally. When changes occur after an event has been computed, the tool needs to allow for recalculation to reveal new results.

*Access*. This option calculates basic object intervisibility from one object to another.

*Deck Access*. This device evaluates from an object to one or more objects found within a database and evaluates if intervisibility is possible within a given the time interval and the look-angle.

*Chains*. Chains calculate object intervisibility.

*Vector Geometry Tool*. The Vector Geometry Tool allows vector geometry to be analyzed and visualized on an object.

*AdvCAT*. This option analyzes possible conjunctions of one or more objects.

*Comm Systems*. Comm Systems evaluates signal strength and impedances.

*Coverage Definitions*. These definitions calculate qualitative evaluations on regions of interests, points, or object paths.

*Grid Inspection Tool*. This tool works with the Coverage tool and refines the analysis within the coverage grid.

Additional tools can be found in the Attached Objects section. These tools need a parent object to attach to and use for analysis. These tools use the parent object's properties and constraints.

*Figures of Merit*. The Figures of Merit tool computes specific statistical evaluations and visualizes the computations.

*Attitude Coverage*. Attitude coverage calculates an object's attitude sphere for qualitative evaluations.

*Attitude Figure of Merits*. From the Attitude Coverage, this tool computes specific statistical evaluations and visualizes the computations.

TABLE 2.2

Tool's Relationships with Objects

| TOOLS | Aircraft | Area Target | Facility | Ground Vehicle | Launch Vehicle | Line Target | MTO | Missile | Planet | Satellite | Ship | Star | Target |
|---|---|---|---|---|---|---|---|---|---|---|---|---|---|
| Access | X | X | X | X | X | X | X | X | X | X | X | X | X |
| Deck Access | X | X | X | X | X | X | X | X | X | X | X | X | X |
| Object Coverage | X | X | X | X | X | X | X | X | X | X | X | X | X |
| VGT | X | X | X | X | X | X | X | X | X | X | X | X | X |
| B-Place Template | | | | | | | | | | X | | | |
| Attitude Simulator Tool | | | | | | | | | | X | | | |
| Export Initial State | | | | | | | | | | X | | | |
| Load Propagation Definition | X | | | X | X | | | X | | X | X | | |
| Close Approach | | | | | X | | | X | | X | | | |
| Lifetime | | | | | | | | | | X | | | |
| Orbit Wizard | | | | | | | | | | X | | | |
| Generate TLE | | | | | | | | | | X | | | |
| Walker | | | | | | | | | | X | | | |
| Solar Panel | X | | | | X | | | | | X | | | |
| Area | X | | X | X | X | | | X | | X | X | | X |
| Laser CAT | X | | X | X | | | | | | | X | | |
| Radio Frequency Interference | X | | X | X | | | | | | | X | | |

*Note:* Object-level tools are also found within the object right-click menu.

In the study of ontology, only certain tools can be used to create relationships with specific objects. Table 2.2 represents a chart to help you identify what tools may be used to relate with which objects.

## STK Objects, STK Tools, and Ontology

Overall, understanding or conceptualizing the semantic level of STK Objects and STK Tools is the ontology within STK. We use ontology to make sure we are getting the highest desired results needed to complete our analysis. STK allows you to analyze objects and compute the relationships of these objects in a robust and sophisticated manner. Without formalizing a study on ontology, many steps to analysis are missed and GIGO becomes a real issue. Applying ontology within the STK environment and refining your STK Objects and STK Tools will help you analyze your projects more effectively and efficiently.

# 3

## The Scenario

### Objectives of This Chapter

- Starting a Scenario
- Planning a Scenario
- Basic Property Information for Objects
- Refining a Scenario by Understanding Constraint Defaults
- Reporting and Analyzing the Information
- Reviewing Some Best Practices

### The Scenario

The scenario within STK is an encapsulated, defined analysis instance of the STK software. The Scenario Level is where global settings and initial analysis time intervals are determined. In order to establish analytical behavior, the environment of where and when the analysis takes place is developed. Plug-ins, add-on modules, and environmental models enhance the behavior of the software from a global perspective. In other words, the parameters that are defined within the global settings have an overall dynamic effect on all the objects and tools that are initiated within the software for localized evaluations. It is important to start off with correctly defined global settings to protect the integrity of the global and local behavior within the software.

### Starting a Scenario

To start a scenario, the user will need to select the "Mountain with a Sunrise" icon. This is found in three primary areas: (1) the Welcome to STK wizard, by selecting the icon that says "Create a New Scenario" button; (2) from the

Task Icon menu section, by selecting the same; or (3) from the Task Menu bar under "File/New. . . ." A user can only run one instance, or scenario, at a time within STK. For best practices, if you need to create several scenarios, then this is handled by creating one scenario, making sure it is saved in a unique file folder, and exiting out of the scenario. After that, a new scenario can be started. Remember, you can only work one instance of a scenario at a time. It is possible to run multiple versions of the software at the same time but not utilizing the same scenario folder. For instance, I often run version 9 and version 10 at the same time from different folders to compare/contrast behavior changes. I use this method often to assist with validation processes.

Visualization options are defined within the 3D Graphics windows, 3D Objects, and 3D Models. With STK software, this is often referred to as VO. There are rumors within the software company on how this term came about. Although the story is irrelevant to the use of the term, it does give insight into the humor and creativity of the programmers and developers within the walls of AGI.

The story was told and handed down in the inner halls of AGI, even though it may not be true, about how VO got its name. Early in the development of STK, the initial visualizations were just 2D window displays. One day, an important client (perhaps a squirrelly U.S. government agency) asked if another visualization option allowing a globe and 3D graphics could be created for the software. Not to be outsmarted by the clients or not accomplish a goal for meeting the client's needs, AGI created the visualization option that allowed 3D capabilities in record time. Thus, the VO was born. Being literal computer science programmers, they didn't know what else to call it, so it was named the *Visualization Option* window, or VO for short.

## Defining Your Scenario

A scenario is an instance of STK ontology, or the semantic study of object relationships, and tasking used for pre-mission, post-mission, or even during-mission analysis. As a scenario task is presented, the user begins to develop a plan of action to create the types of answers to the specific type of tasking that is presented. To create a good plan of action, use the six steps to creating a scenario presented in this section. These six steps will walk the user through the tricks of learning how to begin a scenario, organize the object, and create object property development. A personal tip is to have a notebook

Six Ways to Build a Scenario

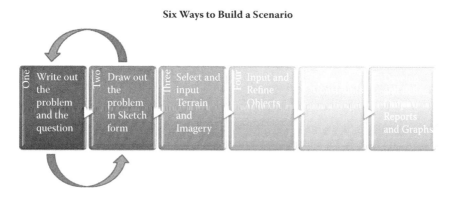

FIGURE 3.1
How to build a scenario.

of scenario-building project logs. This will help the user develop tasking and be a good reference guide.

## Step 1: Understand the Problem and the Question

STK software was designed for modeling, analyzing, and visualizing answers to complex problems. In order to begin using the software, it is very helpful to understand what form of analysis the user will need. Usually, this is in the form of a "Tool" question. By writing out the information needed, it will become easy to see which tool or tools need to be applied to create the solutions needed for your analysis. If clarity to the scenario cannot be realized quickly by just looking at the problem, consider jumping to Step 2—drawing the problem out in sketch form to fully understand the problem.

STK answers questions that are based on spatial-temporal analytics:

1. Where is my object during a determined time interval?
2. When can I see my object from my position?
3. When can other people see my objects?
4. Do I have signal?
5. What is the quality of my signal?
6. Could there be interferences or impedances?
7. How close are my objects?
8. Will my objects collide?
9. What will my orbit transfer look like?
10. How can I plan an orbit?
11. What will my Battle Space Management mission look like?

The questions you ask STK are all tool based. When we begin to clarify the question from a global perspective, we begin to clarify the objects and tools we need to create our scenario.

## Step 2: Draw Out the Problem in Sketch Form

A back-of-the-envelope sketch is highly recommended for scenarios, much like story-boarding is an initial stage to moviemaking. Being able to sketch the problem helps the user understand which objects need to be identified and refined within the scenario and what tools should be applied. The sketch gives the user a better high-level understanding of the complexity of the problem.

The sketch shown in Figure 3.2 tells the user what clues were needed to complete Step 1. The user can now see, for example, that there are communications links that need to be analyzed from a mobile ground vehicle to a ship. Communications would need to be evaluated both for sending and receiving from all objects. The tools relationship would require we use Chains to find the basic Access time intervals. If we need to actually find out when these satellites would be able to communicate with each other, we would need to add in the communication equipment and then evaluate for communication quality after that. A back-of-the-envelope sketch provides valuable information for the STK user and helps prompt questions and further ideas for analysis. Now, the answers to Steps 1 and 2 have been created by sketching out the problem. Steps 1 and 2 may be completely interchangeable if needed (see Figure 3.2).

## Step 3: Select and Input Terrain and Imagery

Inserting terrain and imagery into STK involves both global and local properties. It also has an analytical aspect to it and a visual aspect to the setup. Both are needed; each is handled differently. First of all, analytically, the terrain needs to be introduced as a root-level global property modification.

**Back of the Envelope Sketch**

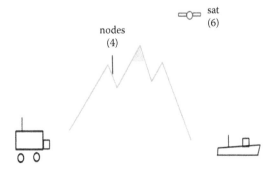

FIGURE 3.2
Back-of-the-envelope sketch.

Analytical and Visual Differences for Using Terrain

| Analytical | Visual |
|---|---|
| • Sample File Types: MUSE Raster, DTED, DEM, GEODAS, MOLA ... PTDD <br><br> • Insert into Root Level of Scenario | • Must Use AGI file type: PDTT or JPeg <br><br> • Covert using Terrain or Imagery Converter (V 9.h) <br><br> • Use Globe Manager for input in 3D to visualize <br><br> • Insert into 2D and 3D Window Properties |
| Used Analytically fir AzEl Masks and other analytica evaluations without the need of making visual for the viewer | Allows for Terrain Visualization |

**FIGURE 3.3**
Analytical and visual differences in terrain.

Inside the object browser, the first object has the mountain with the sunrise icon and the scenario name next to it. This is the root of the scenario. To open the root, right-click on the scenario name and click "Properties." To input the terrain for analytical purposes, use the Basic/Terrain page. To make a file that is not an indigenous STK *.pdtt file into a *.pdtt, the file needs to be converted. This allows the file to be used visually within the software. After the file has been converted, it may be used within the 3D window by using the Globe Manager (see Figure 3.3).

## Step 4. Input and Refine Objects

Review the sketch in Step 2. Notice, that there are objects shown on the sketch. This is the initial inventory list used to begin to build objects into the scenario. As objects are built into the scenario, refining the objects by tweaking the properties to best match real life will give more realistic output. When STK tasking is assigned, it is a good idea to understand the type of fidelity expected for the output. The user is in control of how much detail is put into or left out of the scenario. There may be times when a high-level analysis is all that is needed and other times robust and fully realistic analysis is the only acceptable answer. Usually, this information is given when the tasking is given; however, some of the information may take a little research.

To find equipment specifications, review the Interface Control Documents (ICD) or perhaps the Requirements Document (RD) that has been attached to your specific tasking. Other places to find equipment information would be the manufacturer of the product. The International Telecommunication Union (ITU) has a repository of satellite communications information sold to qualified clients on a subscription basis.

## Step 5. Evaluate Constraints

In STK software, constraints have default parameters for each object and within each tool. This is far different than using the objects within the components building function. Objects built with the components libraries need all the parameters of the objects built into them. It is advisable to thoroughly know and understand what you need to build an object with components; learn the default parameters of the STK Object so you can build the object out. In addition to this, to properly understand how your objects behave and how the algorithms are calculated, it is beneficial to understand what the default parameters are for each object used within STK. In the Objects section, default parameters for each object are defined for STK 9.2.2 based on the object and the Constraint page for the object.

## Step 6. Develop and Refine Output, Reports, and Graphs

In addition to cartographic or movie-simulated output, STK has strong output platforms for basic reports and graphs, link budgets, and customized data providers for graphs and reports. As the objects are defined and then refined by adding constraints and specific properties, the fidelity of the reports and graphs refines to become more realistic. Reports and graphs may be exported; many may be used dynamically on the 3D graphics window. With the use of data providers, custom graphs and reports may be modified to answer the specific questions within the initial framework of the scenario. As objects change over time, STK can evaluate the positional and relationship changes that occur. The reports and graphs can capture these changes for solid documentation.

## Putting It All Together: Creating a Scenario

Creating a scenario is a basic step-by-step process. It applies the ontology within the STK environment. With this set of knowledge, STK becomes a software tool that leverages your strengths as a student, engineer, analyst, or astrophysicist. The body of knowledge from this book should assist you in becoming an STK expert and will make full use of the STK training, science education, and work application easier.

It is recommended to use this book to further your knowledge of STK software on how it works as a Spatial Temporal Information System. This book is not designed to replace the knowledge content that is presented within the training courses offered by AGI. To learn more about actually creating a scenario and the orientation of the graphical user interface within STK, it is recommended to participate in some formal STK educational training prior to delving more deeply into this book.

# Section II

# STK Objects

# 4

## The STK Object

Objectives of This Chapter

- Define the STK Object
- Dynamic Interaction within the Object
- Using a GIS Shapefile within STK
- The Data Federate
- Attribute Page Definitions

## The STK Object

The STK Object models the position, motion, and time interval of real-life objects. These objects may be static or they may move. However, all objects within the STK environment use both position and time intervals in order to be used analytically or visually within the software. The analytical and visual natures require spatial and temporal attributes to relate dynamically to each other. When applying this within the ontological relationship, STK Tools with event algorithms develop unique and interesting spatial relationships.

Spatial analytical systems use similar geometric features within their environment. As a form of Spatial Temporal Information Systems, STK Objects also consist of these three basic geometric feature types: points, lines, and polygons. The point objects represent objects that are either static or propagate. Point objects that are static represent facilities, targets, or places. Objects that propagate, or moving objects, represent forms of vehicles or celestial bodies. Within STK, they are modeled as aircraft, ground vehicles, ships, satellites, missiles, launch vehicles, and planets. Each vehicle type can show and calculate predicted motion or propagation. Area targets are polygons and represent a region of interest. Most STK Objects include import capabilities from the Geographical Information System (GIS) shapefiles into STK. Although not all of the attributal data are transferred from the GIS shapefile,

Feature Geometry of the STK Object

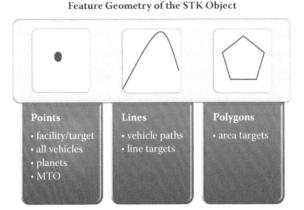

**FIGURE 4.1**
Feature geometry.

the position of the object and an attribute are transferred so STK customization of the object can ensue. Since GIS is database driven and uses time features within the database attributes, the object's conversion is handled in the strictest form and handles the location of the object best. The STK Object models the object spatially and temporally using modeling and simulation techniques where time and attributes are defined within the model properties and not within a database structure.

## Dynamic Interaction within the STK Object

From the previous chapters, we learned that there are two types of STK Objects: parent objects and child objects. All parent STK Objects have some modifiable attributes in common. The STK Object is really broken down into three primary attribute categories that are defined within the properties of the object on a semantic level: Geometric, Time, and Calculations. Within the properties pages, they all consist of four main page segments: Basic, 2D Graphic, 3D Graphic, and Constraint properties. These allow you to fully develop the robust nature of the STK Object type on a semantic level. Child objects inherit, by default, many common features from their parents. They may also be customized by their own property pages.

### Feature Geometry

The Feature Geometry for an STK Object is a point, line, or polygon. The geometry is embedded into the attributes of each object that make the object unique

Dynamic Interaction of the STK Object

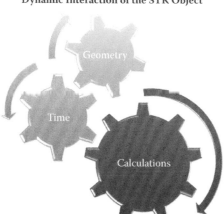

**FIGURE 4.2**
Dynamic interaction.

in physics modeling of motion and object interaction. This motion and object interaction is also known as the kinematics and dynamics of an object. Points are either considered stationary points that represent cities, facilities, or network locations, or they may be vehicle objects that display kinematics during a specific time interval. Lines represent the propagation path from vehicles or line targets. Polygons are created with the Area Target object. The Area Target represents regions of interest, borders of countries, states, focalized areas, or specialized ellipses that are all created with the STK Object Area Target (see Figure 4.2).

## Time Features of the Object

Time is locally defined within the Basic Properties pages of the object instance. This is essential not only for visualization thresholds but also to establish propagation parameters with time steps used to compute the motion of the object, if any. The time is defined by a start time and a stop time. The change in motion also includes the change in orientation and possible velocity.

## Calculations within the Object

Obviously, to identify a change in position, the object geometry and time interval must be considered as well as the object's initial state. As the object's time progresses, the object calculates the next position based on the propagator selected, the terrain or coordinate frame used as reference points to the position, and the environmental parameters. What makes STK so powerful is the dynamic interaction that is possible within the object based on the attributes of the object, the scenario's environment, and the tools used within the ontological study.

## Using a GIS Shapefile in STK

The STK Object instance may import Geographical Information System (GIS) shapefiles within the STK environment based on standard feature geometry types of points, lines, or polygons. Commercial and Open Source GIS software creates shapefiles as spatially referenced vector data. A shapefile requires a minimum of three file types to work as a system: a geometry file (*.shp), the shape index file (*.shx), and an attribute file (*.dbf). In addition to the standard files, STK may also use projection files (*.prj) and spatial index files (*.sbn/*.sbx), or the XML shapefile (*.shp.xml) format to take advantage of the Microsoft.NET interoperability capabilities. There are other file formats used for shapefiles but these are not typically used within the scope of STK, as they relate more to the GIS environment.

Within a GIS environment, GIS shapefiles have a format in a table of records for attribute development within the systems of files that are required to work together. These records do not recognize topology within STK and are primarily used to bring spatially referenced objects into the STK scenario. STK has created a set of GIS tools to import and export GIS shapefiles or overlays. The GIS import tool recognizes the STK Object type it will convert. After the objects are brought into the STK software, each object will need to have the attributes fully developed in the same manner as you would a typical STK Object. For U.S. customers, OILSTOCK overlays are imported using the GIS tool to bring in lines, text, markers, and polygons.

The GIS and STK Objects, although different in structure, are both geospatial objects with built-in positions and shapes. For the GIS and the GIS geodatabase environment, the use of object-relationship database methodology and GIS shapefiles are robust with spatial features. The shapefiles may be created by a number of sources ranging from commercial to Open Source software or freeware.

As the GIS environment is inherently database driven, the format for the shapefile allows relationships to be built using the shapefile primitive data attributes, geometric shape, and geographical representation. However, while the feature geometry, positional information, and a label attribute may be used, at this time the STK environment does not use the shapefile as an object that may be used in a relational database management system (RDBMS) the same way a GIS does. The leveraging of the GIS object within STK allows interoperability between the two spatial information systems.

For GIS and STK, semantic development of the object types and the relationship development, along with their properties, are inherently different. This means that the overall ontological development of the software systems express themselves differently from each other. GIS is inherently database driven. It handles time features as an attribute within the database. On the other side, STK is more object oriented and driven by time and physics. The use of time as a primary function within the software, both globally and locally within the scenario, makes the development of the object and the tools semantically different from the standard GIS object.

Within STK, the GIS shapefile is used to find the spatial referencing, feature geometry, and information from the files to create a label. After the information is selected from the files, STK converts the shapefile to a predesignated STK Object type that matches the feature geometry it is converting it to. For instance, if a user wanted to bring in the outline of a country, such as Belgium, into the STK environment, then he or she would select the Area Target as an object choice. Both the GIS feature geometry of the country and the STK Object types are polygons. When the GIS shapefile is brought into the STK environment as an Area Target, the file name of the object would no longer be Belgium.shp but would convert to an STK Object type and would be Belgium.at—the *.at format is the extension used for area targets. Once the shapefile has been converted to a useable STK Object type, the STK properties and attributes may then be defined.

STK has an archived library of GIS shapefiles that are currently nested within the program files for the software. For version 9.*n*, these can be found in C:\Program Files (x86)\AGI\STK 9\Data\Shapefiles if using a 64-bit machine or C:\Program Files\AGI\STK 9\Data\Shapefiles for a 32-bit machine. The location of this file may vary with different versions of software and the hardware an individual is using. There are four main categories of shapefiles found in the program files within this folder for shapefiles: Area of Regards (AOR), Countries, Land, and Water. All of these are polygons. However, this does not limit the shapefiles that may be brought into STK. As long as the feature geometry shape is met so that the conversion may take place—vehicles (points), cities (points), and other shapefiles may be brought in from a GIS repository.

---

## Using the Data Federate

The STK AGI Data Federate® (ADF) is a data storage management system designed to store STK Object models, scenarios, and even STK output results. The data storage may be used as a centralized repository system to allow multiple users access to use, create, modify, and share data across teams or business centers. Domain permissions allow for a hierarchy of data permissions. With domain permissions and versioning control, a user is able to trace and track all edits, users, and modifications. This allows strong quality assurance and quality control (QAQC) over data. A data administrator has full control over the stored data, allowing read, write, edit, and copy privileges for each object within the ADF. For developers, STK has class-level and application program interface (API) capabilities to allow integration with non-AGI software such as GIS software, SharePoint®, Teamcenter®, and other third-party products.

## The STK Object Instance

Every instance of STK Object has four common property sections: Basic, 2D Graphic properties, 3D Graphic properties, and Constraints. Each common property section has a listing of pages that have modifiable attributes within the pages. Although some of these pages have common features, some are different based on the type of object and the feature geometry. To fully refine the object instance, it is best to start defining the basic properties and systematically work through the pages consecutively. When a user has finished working down through all the property pages within the 2D Graphics, 3D Graphics, and Constraints, then the object is fully defined. After defining the object, modifying the definitions becomes easy (see Figure 4.3).

### Basic

Basic properties define the position, velocity, propagation, trajectory method, and/or the area of the object. The pages available are dependent on the object type. For point objects, such as facilities and targets, these objects represent one position with a latitude and longitude, and many times altitude. Point mass vehicles, such as ground vehicles, satellites, missiles, ships, and aircraft, also have velocity information included. This means a point mass vehicle may have many positional points to evaluate. The simplest form of this is by creating an initial position for the vehicle that includes orientation and spatial position and then adding time intervals where these positions or orientations might change. This motion is handled by a variety of methods, primarily using robust physics-based propagators that are designed to handle the specific calculations for the correct object type. Examples include the Aircraft object that uses a sophisticated Aircraft Mission Modeler (AMM) for modeling flight

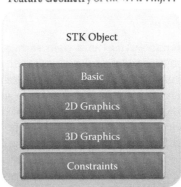

FIGURE 4.3
Object instance.

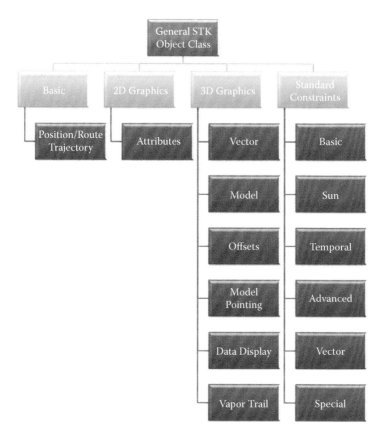

**FIGURE 4.4**
Object hierarchy.

dynamics or a satellite that uses the Astrogator propagation model. Polygon object types are Area Targets that consider a region of interest made up of multiple positional points and create a closed polygon (see Figure 4.4).

## Basic Attributes in Common

Based on the STK Object type, there are some basic attributes in common. Primarily, when working with the point mass objects, these include the Great Arc Propagator and the Attitude attribute pages, which are similar in nature.

### *The Great Arc Propagator*

Some point mass objects, such as a ship, ground vehicle, or aircraft with vector capabilities, use the Basic property "Great Arc" for the route propagation. This calculation gives interpolated results using the great-circle distance algorithm starting with an initial position the user has defined and then calculating the based on bearing and waypoints defined. User-defined attributes that define the Great Arc are:

**Time Interval:** The standard default time interval is the full global scenario time frame. This may be customized by deselecting the "Use Scenario Start Time" and defining the time you choose to use.

**Altitude Reference:** Altitude Reference allows three options: WGS84 (default), Mean Sea Level (MSL), or Terrain. Using interpolated results, object references are used for different reasons based on the level of smoothness. The smoothest algorithm is the MSL. The WGS84 is less smooth than MSL but much smoother than the use of Terrain files. Terrain is the least smooth. The smoothing is based on the intervals of vertical Altitude References.

Altitude Reference is the vertical reference of the waypoint of the object from the reference geoid/terrain type. For instance, if you change the altitude of the waypoint and set the Altitude Reference to Terrain, the object will maintain that altitude above the terrain and follow the arc paths above the terrain height values. The analysis is interpolated to decrease computation cost.

**Route Calculations:** Smooth is the default parameter. You may enter Rate and Acceleration values, or the time entered for each waypoint.

**Waypoints:** These may be inserted within the Properties window or you may input them by clicking on the 2D Graphics window with the Basic Route properties open. Waypoints, if defined using a table format within the properties, allow you to define the latitude, longitude, altitude, speed, acceleration, and time the vehicle is at the waypoint, as well as the turn radius of the vehicle.

**Arc Granularity:** This is also user defined with the default granularity for calculations set at 0.572958 degrees.

The Great Arc file is an ACSII text file that ends in a *.ga or *.pa extension. This file may be modified if needed as long as the file format is left intact and imported into the software for your use. A copy of this file format is found in the Help files for use as a template.

### Position: WSG84, On Terrain, or MSL

When we look at the Earth's model in the figure, we note there are differences between elevation points and smoothing effects. This is a hand-drawn sketch to emphasize the differences between what happens within the STK software when it references points from the WGS84, Terrain, or Mean Sea Level options. The World Geodetic System (WGS84) is a survey of both vertical (latitude and longitude) values and horizontal (altitude) values, along

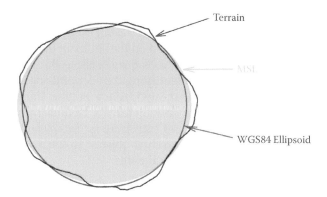

**FIGURE 4.5**
The Earth's ellipsoidal shape is represented as an oblate spheroid in this figure.

with measurements of the mean sea levels, gravitational changes, and geoid measurements. The WGS data is published by National Geospatial-Intelligence Agency (NGA). This model represents the Earth's ellipsoidal shape as an oblate spheroid. WGS84 is used for the geodetic coordinate system values within STK. This is the default geoid definition used for Earth's central body calculations of position.

Mean Sea Level is an option to use for geoid measurement values within STK. This is an evaluation of averaged tides and waves over various positions of the Earth and over a period of geological time. The vertical means of these measurements are observed hourly over 19-year lunar cycles. The values of MSL are smoother than the topological surface found when we add in terrain, but are more vertically defined than the smoother algorithm of just using the WGS84 ellipsoid. MSL is a better algorithm for aircraft flight. Mean Sea Level can only be used for Earth's central body.

Terrain can be used by any central body that has the terrain information. These topological maps model truer-to-life horizontal and vertical changes than the smoother algorithms, the WGS84 ellipsoid, or the MSL. Terrain in STK may be used analytically in many formats. However, if the terrain needs to be visualized within the software, it must be converted into a PDTT format using the Terrain or Imagery Converter tools.

### The AzEl or Body Masks

An Azimuth-Elevation Mask is an area of visibility defined by the azimuth angle combined with the elevation angle. We define an azimuth as the horizontal angle of a specified object from a point or origin. An elevation angle is an angle that measures horizontally to a specific object. Azimuth-Elevation Masks, or AzEl Masks, limit line-of-sight calculations to an object. Just like many other things within STK, the AzEl Mask is both analytical and visual. The analytical attribute is created in the Basic/AzEl Mask page, made visual from the 2D Graphics/AzEl Mask page, and then further refined within the 3D Graphics/AzEl Masking page. The mask file (*.aem) may be selected based on terrain or may be custom developed based on the ability of the equipment and obstruction points. Another term for this is *Body Masking*. Body Masking files (*.bmsk) may be created in STK by identifying the six points needed to represent the obstruction or they may be created from a graphic image. The AzEl Masks or Body Masks are frequently used in targets, facilities, and some child objects, such as sensors, receivers, and transmitters. Samples of these mask files will be presented in Chapter 11 on child objects. To learn more about the AzEl Mask, refer to the STK training documents.

## 2D Graphics

2D Graphics properties are visual attributes of the object that define the mapping details. This includes attributes of propagation lines, font, lighting, display times, and contours. This allows you to refine the object to create a higher-quality visual appeal for all forms of output—the visual analysis, the animation video, or static maps.

For moving objects, the Attributes page allows you to be able to visualize the route path by selecting the properties of "Show Label," "Show Route," and "Show Route Marker," all toggled on by default. At the top of the page, there is the ability to modify this visualization to color-code the path, for example, when an object has "Access" or intervisibility between this instance of the object and another selected object or group of objects. A user also has the ability to select custom intervals.

Other vehicle graphics are visualization aids to help us understand more about the position and the view of the object we are modeling. Elevation and Range Contours allow you to see contour graphics at either an elevation angle or a specified range value. The Vehicle Lighting defines the way light is displayed around the object. The Vehicle Swath calculates with several swath choices to choose from: Ground Elevation (default set at zero [0] degrees), Vehicle Half Angle, Ground Elevation Envelope, and Vehicle Half-Angle Envelope.

## 3D Graphics

3D Graphics properties may directly affect the computational values of the objects as relationships are established. These changes also affect visualization of the objects within the 3D window environment (see Figure 4.6).

### Vector Settings

The Vector Settings page allows you to apply, modify, and visualize vector geometry in regard to the STK Object. Additionally, this page is one of many avenues to open up the Vector Geometry Tool (VGT). Like many

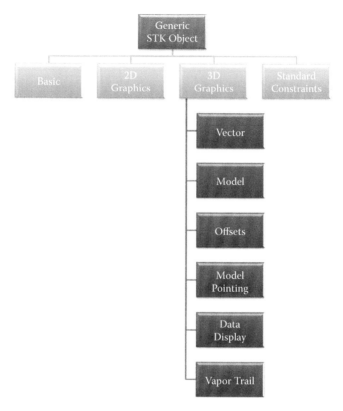

**FIGURE 4.6**
3D graphics hierarchy.

tools within the STK environment, the graphics are created first and then visualized. It is a two-step process. There is an array of already created VGTs or you can customize your own using the VGT features. After the vector geometry has been created, you need to visualize it by defining its attributes within the STK object in the 3D Graphics Vector page or within the 3D Graphics window.

For static objects, the vector geometry is referenced from the plane within where it is located. For instance, if we used a Facility object on Earth, we would use the standard visualization. We could do the same for a moving object—albeit, because a moving object is a point mass, the vector geometry will include more parameters to determine orientation, motion direction, and velocity, all within a time interval selected.

While many of the STK vectors are well defined, you can modify and create your own custom vectors for your object by using the Vector Geometry Tool (VGT). The VGT is a nonontological tool designed within the environment of STK and is discussed in the Tools section of this book (see Figure 4.7).

**FIGURE 4.7**
VG facilty.

**FIGURE 4.8**
VG plane.

## Model Settings

The 3D Graphic Model page is a common page for most STK Objects; one exception is the Area Target. This page is where you change the object model. You can visualize, set the object to create motion (transformations) within the Articulations page, and change the attach points for the object. In STK, this is handled within the 3D Model Properties page for the STK Object. To give the object motion during animation, where the propellers move on an aircraft or the radar dish slews to maintain pointing toward a specific targeted satellite, for example, modification to the object's Articulation file is required.

## The STK Object Model

The STK Object Model is a 3D mesh-model visual representation of a real-world entity. These models allow you to visualize the motion (kinematics) and the dynamics (relationships) of the objects during the interval of time. The early 3D model used the MDL format, which uses traditional mesh model computer-aided design (CAD) methods with text file buildups and a model articulation file to control the kinematics. The MDL (*.mdl) format is still supported by STK. STK recently adopted support for the COLLADA™ (*.dae) file. Additionally, AGI has created a utility to convert the Lightwave (*.lwo) model to a working MDL file format. You may download this utility from AGI.com. Keyhole Mark-up Language (KML) files from Google may be used visually within the software but at this time do not carry any analytical data with them and are not fully supported.

The comprehensive training courses have excellent support in showing you how to create motion using the STK Object Model. The courses discuss how to create a Model Articulation file (*.ma) using manual techniques or using the Timeline tool. Both give STK a more realistic visualization to the scenarios.

## 3D Graphic—Offsets Settings

The default parameter for most STK Objects is to have all visualizations and computations based on the centroid of the object. However, on the Offsets Settings page, this can be modified visually. Visual offsets for objects may be established that make the object appear to be in a slightly different position from the calculation point to make the visualization items appear more realistic. It is important to note that the calculation point will not be moved. There are three positions to offset an object that may be used simultaneously: Rotational (*x*, *y*, *z* body-axis positions), Translational (*x*, *y*, *z* body-axis positions), and Label Offset, which modifies the position of the object's label.

The Attach Point option allows you to select a model articulation for Attach Point referencing for display lines. You would use this to direct the

Access visualization and cone swaths to appear as if they were coming from the center point of the articulation versus the center point of the STK Object's main body.

### 3D Graphic—Model Pointing Settings

The Model Pointing page allows you to point an object to always face a specified object, target, sun, or central body. It uses the analytics of vector geometry to assign objects for your object to point toward.

### 3D Graphic—Data Display Settings

Data Display of the object, such as Latitude, Longitude, Altitude (LLA) or Velocity Heading readouts for the object may be dynamically displayed on the 3D window so the user can see how these readouts change over time.

### 3D Graphic—Vapor Trail Settings

If the object is able to display motion, vapor trails may be graphically displayed on the object to increase visual effect. The effect of showing gas as it escapes the tailpipe is a strong visualization tool.

## Constraints

The Constraints section's standard basic properties are universal to most objects. Primarily, most objects have constraint pages entitled Basic, Sun, Temporal, Vector, and Special. In Figure 4.9, we note the different pages of constraint types found within the Constraint's section of the generic STK Object. The Constraints Basic page and the Sun page allow the user to see the default parameters already attached to this object and modify these parameters when needed. Temporal and Vector constraints allow special modification of time intervals and to constrain a vector angle or magnitude. Special constraints are object specific, such as terrain grazing angles or specific object exclusions within ontological relationships as they refer to this specific object.

Standard constraint default parameters are important to understand. The entire ontology of the objects considers the settings within the constraints. One way to refine your calculations is to define the parameters of the constraints to meet your requirements. These default parameters are not seen unless you toggle them on by selecting the minimum/maximum settings. All of these parameters may be modified as needed. Not all objects will use all categories within the constraints pages, and a user cannot modify the constraints that are not used for these objects. The object constraints you do not have access to should be a lighter color gray on the page you are looking at. The Help files define all the terms of each page visually and semantically.

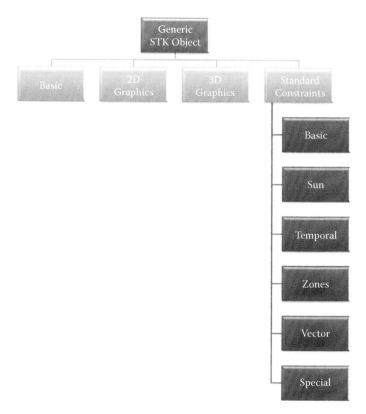

**FIGURE 4.9**
Constraints hierarchy.

### The Constraints—Basic Default Settings

The Constraints Basic page restricts the event-detection calculations from the perspective of the object during the event evaluation. All the settings include the time intervals by default. However, you can exclude time interval calculations for each specific basic constraint if desired by toggling it off. The Line of Sight (LOS) constraint, which is on by default, shows the object does not have a successful Access event if the Line of Sight of the object to another is obstructed. Both the Azimuth-Elevation Mask (AzEl Mask) and Terrain constraints are off by default—these are used primarily with the Target or Facility STK Objects. When the AzEl Mask constraint is used, best practices state the Line of Sight constraint is also turned on for correct calculations.

### The Constraints—Sun Default Settings

The Constraints Sun page allows you to constrain your object based on the relative position of the sun or moon to the STK Object (see Figure 4.10). The

**The STK Object's Constraints Basic Page**

| Azimuth Angle | Elevation Angle | Range |
|---|---|---|
| • Min: 0 deg<br>• Max: 360 deg | • Min: 0 deg<br>• Max: 90 deg | • Min: 0 km<br>• Max: 40,000 km |

| Azimuth Rate | Elevation Rate | Range Rate |
|---|---|---|
| • Min: −5 deg/sec<br>• Max: 5 deg/sec | • Min: −5 deg/sec<br>• Max: 5 deg/sec | • Min: −5 km/sec<br>• Max: 5 km/sec |

| Angular Rate | Altitude | Propagation Delay |
|---|---|---|
| • Min: 0 deg/sec<br>• Max: 5 deg/sec | • Min: 0 km<br>• Max: 20,000 km | • Min: 0 sec<br>• Max: 3,600 sec |

| Line of Sight | Terrain Mask | Azimuth Elevation Mask |
|---|---|---|
| • On | • Off | • Off |

**FIGURE 4.10**
Constraints basic defaults.

**The STK Object's Constraints Sun Page**

| Sun Elevation Angle | Sun Ground Elevation Angle | Lunar Elevation Angle |
|---|---|---|
| • Min: 0 deg<br>• Max: 90 deg | • Min: 0 deg<br>• Max: 90 deg | • Min: 0 deg<br>• Max: 90 deg |

| Line of Sight | Boresight | Sun Specular Point Exclusion |
|---|---|---|
| • Solar Exclusion Angle: 0 deg<br>• Lunar Exclusion Angle: 0 deg | • Solar Exclusion Angle: 0 deg<br>• Lunar Exclusion Angle: 0 deg | • 0 deg |

| Lighting | Solar/Lunar Obstruction | Additional Central Body Obstruction |
|---|---|---|
| • Direct Sun (default)<br>• Umbra<br>• Penunbra | • Off | • Off |

**FIGURE 4.11**
Constraints sun defaults.

Boresight constraint is only available when a child object is attached. It refers to an attached child object's boresight, such as a sensor, relative to the position of the sun or the moon.

### The Constraints—Temporal Default Settings

The Constraints Temporal page allows you to specify the temporal aspects of the STK Object, based on local and GMT start/stop times, duration, and selected intervals. This constrains the ability for the object to be able to be included or excluded from Access computations. By default, the object may always be considered in the Access calculations—but even *if* Access were possible, there are times that we might need to define that possible object as off limits to the calculations. This tool is used for that reason.

### The Constraints—Zones Default Settings

The Constraints Zones page allows you to select specific zones to either include or exclude from your computations. These zones are based on latitude and longitude regions.

### The Constraints—Vector Default Settings

The Constraints Vector page allows you to constrain the applied Vector Geometry as defined within the 3D Graphics Vector page. There are two possible ways to constrain vectors here: by Angle and by Vector Magnitude.

### The Constraints—Special Default Settings

The Constraints Special page allows you to constrain your STK Object by Geostationary Belt Exclusion, Height above Horizon, Terrain Grazing, or Elevation Rise. You may also exclude a specific object by angle or time.

## Understanding the STK Object

Every STK Object instance is defined by basic parameters. The definition is first defined by selecting the object type and feature geometry. After the object type is selected, the attributes are given by default parameters and then each object may be further refined semantically by modifying the parameters and making them unique to the specific object instance. These objects model the spatial temporal dynamics and kinematics of a real-life object. Now that we have looked into general STK Objects, we can further look into the specific STK Objects and the specialized attributes that define them.

# 5

---

## *Area Targets*

---

### Objectives of This Chapter

- Defining an Area Target
- Methods of the Area Target
- Defining an Area or Ellipse
- 2D Graphics
- 3D Graphics
- Constraints

---

### Area Target Objects

The Area Target object is a region of interest on a central body. The feature geometry for the Area Target is a closed polygon or an ellipse. The closed-system object is often used to allow analytical evaluations in a region. We frequently use this for communications and sensor footprint analysis.

To set up the Area Target, there are many methods to select from within the object: Countries and US States database, the use of an Area Target wizard, a saved shapefile or target file, the AGI Data Federate, or default properties. Additionally, the Area Target object may be modified by using the Area Targets Basic Properties pages or the 3D Graphic editor after the object instance has been developed (see Figures 5.2, 5.3).

#### Methods: Countries, US States, and Shapefiles

Creating Area Target objects using the method of either selecting from the Countries and US States database or by importing shapefiles (*.shp) derives the objects' information from polygons (see Figure 5.4) created within a geographical information system (GIS) or spatial information system (SIS). The file is then converted to a usable *.at file format. Shapefiles are sets of spatial

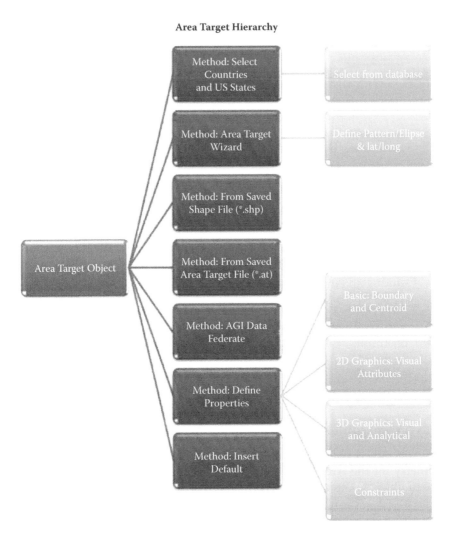

Area Target Hierarchy

**FIGURE 5.1**
Select AT hierarchy.

features with attributes. Data storage for a shapefile is considered small, with easy read, write, and edit capabilities.

You will note that the .xml file format holds the attribute file and the shape index embedded within the file. The .xml format is based on the Federal Geographic Data Committee (FGDC) metadata standards. This includes longitude, latitude, and altitude information as well as coordinate system and other indexed information.

The Countries and US States database has a list of common shapefiles that can be brought in and automatically converted to a usable Area Target file format. However, if the method of "Selecting a Shapefile" is chosen to

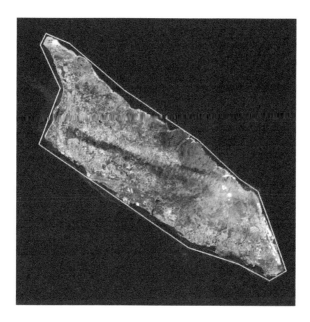

FIGURE 5.2
Aurba AT.

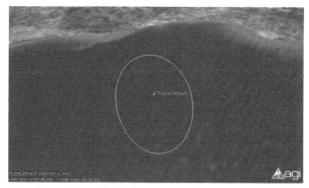

An Area Target Ellipse with default parameters. The Semi-Major axis is 360
km, the Semi-Minor axis is 180 km and the bearing is at 0 degrees. The
centroid will sit at 0 degrees latitude and 0 degrees longitude.

FIGURE 5.3
Ellipsoid AT.

bring in a customized area, then you will need to manually convert the
shapefile to the Area Target file format (*.at). Shapefiles store spatial points
of interest.

Notice the "Name By" section in the "Import Shapefile" method. The pull-
down menu includes the list of items found in the field names located in the
Aruba.dbf file listed in the figure. This is the same information found in the

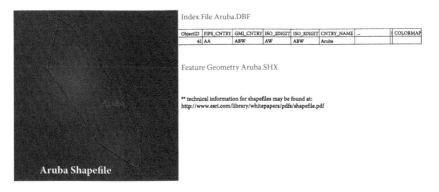

**FIGURE 5.4**
Aruba shapefile.

**FIGURE 5.5**
Importing a shapefile.

Aruba.shp.xml embedded in the file. STK uses the file-name information as a label.

## Method: Area Target Wizard

The Area Target wizard allows you to create both feature geometries of a polygon (also called a pattern) and an ellipse. The wizard assists in the creation of the perimeter for the polygon. In addition, elliptical boundaries are also set with user-defined options by Semi-Major and Semi-Minor Axis, Bearing, and Centroid.

## Defining Properties

### Basic Properties

The Basic Properties of the area target are what define either the polygon pattern or elliptical boundary of the object instance. The center point, or centroid, of the area target is autocomputed for the polygon by default. This is the point where the Access lines are drawn visually during animation. However, the centroid may also be hand-selected by clicking on the 2D Graphic window or by typing in the centroid latitude, longitude, and altitude based on geodetic, spherical, Cartesian, cylindrical, or geocentric position. The elliptical centroid is not autocomputed. The centroid for the ellipse needs to be hand-set (see Figure 5.6).

### 2D Graphics Properties

The 2D Graphic Property Attributes page allows you to establish what color, line style and width, and visual enhancements may be used for the area target. The display times for the object default parameters are set to show as always on. Other settings include visualizing only on during an access event or by defining specific intervals (see Figure 5.7).

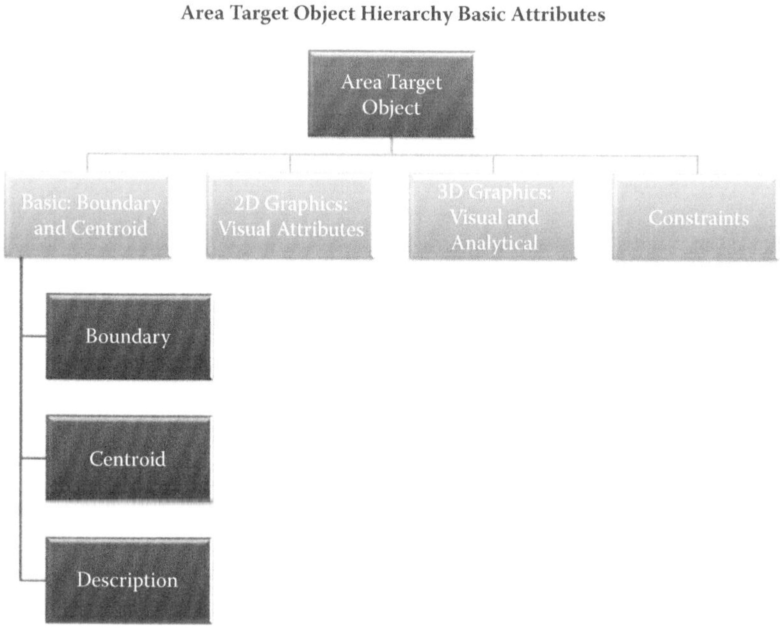

**Area Target Object Hierarchy Basic Attributes**

**FIGURE 5.6**
AT hierarchy.

Area Target Object Hierarchy 2D Graphics Attributes

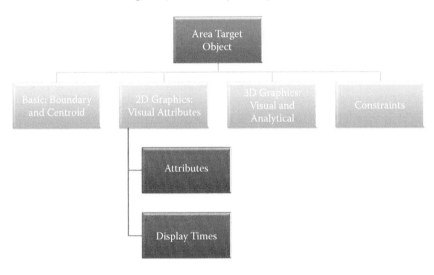

**FIGURE 5.7**
2D AT.

## 3D Graphics Properties

There are two main pages within the 3D Graphics section of the pattern type Area Target object, the Attributes and Vector pages (see Figure 5.8). The Attributes page allows you to create a max viewing distance, fill the interior of the Area Target object with color (based on your choice found within the 2D Graphics Attribute page), and select level of translucency. Lastly, the 3D Graphic Attributes page allows you to create a Boundary Wall using the perimeter boundary line. This line allows you to display both upper and lower edges, which may be set from the World Geodetic System, 1984 edition (WGS84), Height from Terrain option, or the Mean Sea Level (MSL) option.

### Defining The Position of Boundary Line

World Geodetic Survey, 1984 edition (WGS84), is the default position and visualization for the area target boundary line. This position for the boundary line can only be used with the Earth Central Body choice. The Height from Terrain option is available for all central bodies available (see Figures 5.9, 5.10). This option allows you to choose upper and lower boundary vertical positions from the altitude of the terrain file. The Mean Sea Level (MSL) option is only available for Earth Central Body.

**Area Target Object Hierarchy 3D Graphics Attributes**

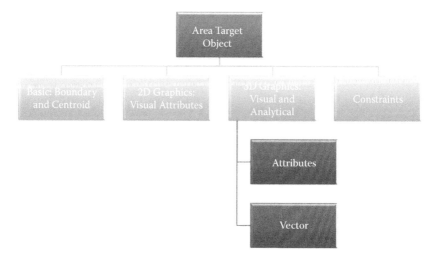

**FIGURE 5.8**
3D AT.

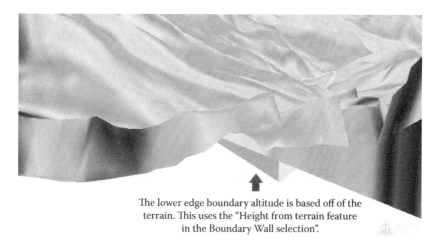

The lower edge boundary altitude is based off of the terrain. This uses the "Height from terrain feature in the Boundary Wall selection".

**FIGURE 5.9**
Boundary line bottom.

## Vector

There are basic default vectors already created and available for your use on the 3D Graphics Vector page. They allow you to display and understand the vector geometry relationships of the Area Target object. You have user options to select the localized central body these objects draw from. This is important if your area target is located on another central body other than

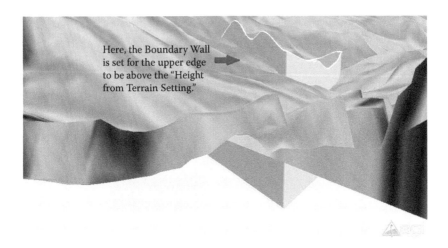

Here, the Boundary Wall is set for the upper edge to be above the "Height from Terrain Setting."

**FIGURE 5.10**
The upper level boundary wall is also determined from the edge of the terrain using the "Height from Terrain" level setting.

Earth. Earth is the default and unless you have changed the central body axes points, it will still continue to draw from Earth's position. After you have verified the central body you are drawing the area target from, you have options to modify the color, persistence, component size, angle size, and arrow size for visualization effects (see Figure 5.11).

## Constraints

Constraints for an area target are slightly different than other objects. The Constraints Basic page has options to modify the minimum elevation point. The default is set at zero (0) degrees. This elevation angle is based on the tangent to the surface of the Earth, which is also known as the local horizontal plane. You may also select the option to compute access for the entire object. Lastly, the default parameter of Line of Sight (LOS) is toggled on and will calculate for LOS analysis unless toggled off.

There are many parts of the STK Object that have identical attributes. The Temporal Constraints are the same as previously discussed in the STK Objects chapter and are used to define that possible object as off limits to the calculations. The Vector constraints are also the same as mentioned before and require applied vectors from the 3D Graphics Vector page to assign values for angle or vector magnitude. After this chapter, unless it is essential to understand the object's functionality, common attributes are not reiterated (see Figure 5.12).

**Area Target Vector Geometry**

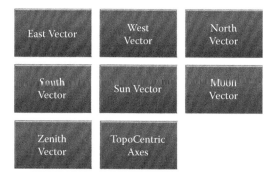

**FIGURE 5.11**
The 3D Graphics/Area Target Vector Geometry Page options allow the user to select vector geometry for 3D visualization. Access to the Vector Geometry tool may also be found on this page.

**Area Target Object Hierarchy Constraint Attributes**

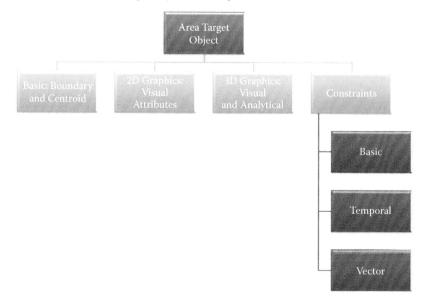

**FIGURE 5.12**
The Area Target's Constraint pages include Basic, Temporal and Vector constraints. These constraints have default parameters and may be modified as needed to make the analysis more realistic.

# 6

## Targets, Facilities, and Places

### Objectives of This Chapter

- Methods to Define a Fixed Point Object
- STK City Database Files
- Facilities Database File

### Fixed Point Objects

When we review positional data, we often need to create a static position to represent a specific location. This location could be a city, an installation, or even just a place. In STK, this object is modeled as three different objects: the Target, Facility, and Place objects. However, these three objects are essentially the same in structure and form. The feature geometry in all is a point feature. Besides the icon that represents the objects, the only real difference is the STK Object extension: the Target is a *.t object, a Facility is a *.f object, and a Place uses a *.plc object extension. The Target, Facility, and Place objects all represent a single fixed point on a central body. They may represent a targeted position based on latitude, longitude, and altitude, or they may represent a city or facility or simply a fixed position in time and space. The methods for the objects are the same and utilize the same databases, files, and properties (see Figure 6.1).

When we use point files, we either use what is already in the STK database files or we customize our own set position within the STK environment. Cities, targets, and places all use the City and Facility database files found with STK. These files can also be customized to create a network of cities or facilities of your own.

**Target/Facility/Place Hierarchy**

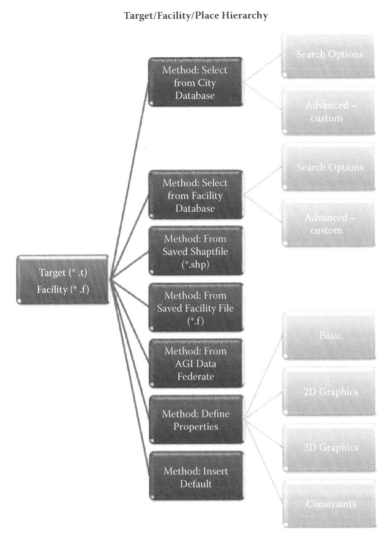

**FIGURE 6.1**
Target hierarchy.

## City Database Files

While most of the information regarding city database files resides in the Help files, not all of it is found in the same page system for easy reference. The following information is designed for you to be able to customize your own databases within STK easily. The STK Help files define the City database

**FIGURE 6.2**
This 2.5D map overlay is a graphic of a city. We can create our own custom city, building, target and place databases within STK for easy access to our own unique objects used for analysis.

as requiring three formats: stkCityDb.cd, which is the main database; stkCityDb.cc, which is designed to hold the Country files; and stkCityDb.gd, which gives current update information. All three of these file formats are necessary to complete the action for City database use. The default STK City database has included a fairly extensive list of cities and countries for general use. However, many customers have found advantages to customizing the City database for individual use.

If you examine the characteristics of the three file formats as they are shown here, you will have the tools to customize your City databases. These files are found at C:\ProgramData\AGI\STK 9\Databases\city in your computer directory browser outside of the software environment. The stkCityDb.cd, stkCityDb.cc, and stkCityDb.gd are all there for you to review in Notepad or a text editor environment.

Column information for the stkCityDb.cd file is also found in the Help files. For convenience, it is reiterated in Table 6.1. The Column section is not represented on the template file as a specific attribute type, but is only used to help the user find the correct column position from the template.

City database stkCityDb.cd file formats are formatted to look like that shown in Figure 6.3. The City database stkCityDb.cd file format is shown in Figure 6.4, and City database stkCityDb.gd file formats are shown in Figure 6.5.

**TABLE 6.1**

City database information for stkCityDb.cd

| Column Section | Attribute Description | Columns Used | Column Positions |
|---|---|---|---|
| 1 | Unique Identifier | 7 | 0–6 |
| 2 | City Name | 30 | 7–36 |
| 3 | City Type (Choose one): 1—Populated Place; 2—Administration Center; 3—National Capital | 2 | 37–38 |
| 4 | Country | 20 | 39–58 |
| 5 | Province/State | 40 | 59–98 |
| 6 | Province Rank | 3 | 99–101 |
| 7 | Population | 11 | 102–112 |
| 8 | Population Rank | 3 | 113–115 |
| 9 | Latitude (deg) | 17 | 116–132 |
| 10 | Longitude (deg) | 17 | 133–149 |
| 11 | Central Body | 12 | 150–162 |

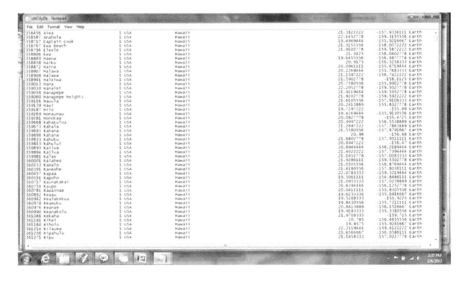

**FIGURE 6.3**
STK City Database stkCityDb.cd file.

## STK Facility Database Files

The Facility database is also customizable. Facilities are networks of specific locations. For instance, you might have a network of radar dishes that you

**FIGURE 6.4**
stkCityDb.gd.

**FIGURE 6.5**
An example list of cities in the stkCItyDatabase.cc.

would like to represent as a single location, or you might want to map all the NASA sites that receive telemetry for your satellite. Whatever your specialized need is, it is possible to create your own Facility database. If you examine the files below, you can set up your own network and use it within STK easily. STK Help files define the Facility database as requiring three formats:

stkFacility.fd for the Primary database file, stkFacility.fn for the Facility network type, and stkFacility.gd for the latest updated information. The facility network file, stkFacility.fn, includes both network types (e.g., NASA STDN) and a selection of central bodies to choose from. Please note that in order for the database files to work, all three must be correctly formatted and present within the correct database directory. The working database directory is C:\ProgramData\AGI\STK 9\Databases\Facility in your computer.

Facility database stkFacility.fd file format is shown in Table 6.2.

An example of a Facility database stkFacility.fd file format is shown in Figure 6.6, and a Facility database stkFacility.fn file format is shown in Figure 6.7. A Facility database stkFacility.gd file format, which describes the latest update information, is shown in Figure 6.8.

TABLE 6.2

Facility database information for stkFacility.fd

| Column Section | Attribute Description | Columns Used | Column Positions |
|:---:|:---|:---:|:---:|
| 1 | Site Name | 37 | 0–36 |
| 2 | Network Type | 12 | 37–48 |
| 3 | Latitude (deg) | 10 | 49–58 |
| 4 | East Longitude | 11 | 59–69 |
| 5 | Altitude | 7 | 70–76 |
| 6 | Central Body | 12 | 78–89 |

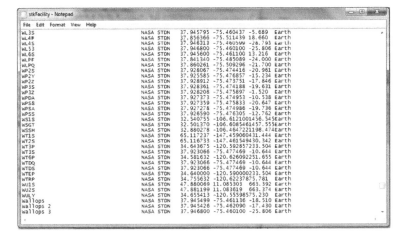

FIGURE 6.6
Facilty.fd file.

**FIGURE 6.7**
Facility.fn.

**FIGURE 6.8**
stkFacility.gd.

## The Static Point

The City, Facility, and Place object is the simplest object within the STK environment. It is a simple static point. It has one geolocation possible at one time. It has a longitude, latitude, and an altitude. The point does not carry motion. It does not have forces that are recognized. There isn't a way to factor propagation. The simple static point is a geospatial position. Next, we will look at the point mass, which is a point that has vector geometry and can model motion.

# 7

## The Moving Object

---

### Objectives of This Chapter

- The Moving Object and Time
- Points, Point Mass, and Vector Geometry
- Moving Object Propagators

---

### Moving Objects

As we have learned from previous chapters, the STK Object's feature geometry is defined as a point, a line, or a polygon. The moving object is a point with vector geometry. In the world of physics and computer science, this specialized point is called the point mass. The vehicle point mass uses vector geometry to build robust analytical models of the propagation, orientation, and behavior of a moving object. Spatial Temporal Information Systems use point mass physics modeling to evaluate an object's motion, velocity, and orientation over time. Vehicle objects include the Ground Vehicle, Ship, Aircraft, Launch Vehicle, Missile, and Satellite. In this book, our focus beyond the basics is on the high-level use of an aircraft and the satellite, as they are the most used objects, with some mention of the launch vehicle and the missile objects. Since these objects all have a point mass structure that includes vector geometry and physics-based properties, the general rules apply to all of the moving vehicles.

Within the parameters of the ontological relationships, the necessity to fully define an object is essential. The moving object is not different in this regard. In fact, because of the object's need to compute predicted motion or propagation defined by the attributes within the object prior to a relationship being established, the ontological details become more refined. Let's take a moment and review the definition of a Spatial Temporal Information System like STK and compare how it is different from a spatial information system in the general domain of spatial analytics (see Figure 7.1).

**Spatial Temporal Information Systems**

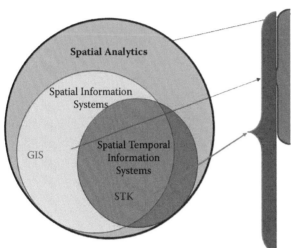

- Understanding position of things and events
- Analyze the generalized mapping details and their relationships on a 2D display
- Use Raster and Vector data
- Create displays of analysis output in cartographic form
- Ability to alter, modify or change map elements

- Analyze congruent and incongruent time intervals for mapping elements in 3D
- Allows for event prediction over time
- Allows for vector geometry analytics overtime with event detection

**FIGURE 7.1**
Spatial Temporal Information Systems.

Spatial analytics is the general study of the positions of objects and events in a coordinate system. When we develop relationships with these objects or events, we begin to deepen our spatial understanding. Patterns and spatial contexts emerge as relationships are refined. Adding a time correlation into the analysis permits you to broaden your understanding, predict motion and events over time, and allow vector geometric attributes to be considered to a greater extent.

## Spatial Temporal Analytics Using a Time Clock

Dr. Waldo Tobler, the renowned "Father of Geography," once made an observation regarding relationships of objects around us. He said, "All things are related, but near things are more related than far." This statement is known as the First Law of Geography. There is a similar law in physics discovered by Sir Isaac Newton. He discovered quantitative or qualitative analysis could be observed with relationships as they interact over time. These relations were dynamically interdependent based on the distances between the objects. The further away the objects in the relationship, the lesser the dynamic effects. Orientation, vector motion, and velocity, as well as mass proportions, also dynamically affect objects in motion (see Figure 7.2).

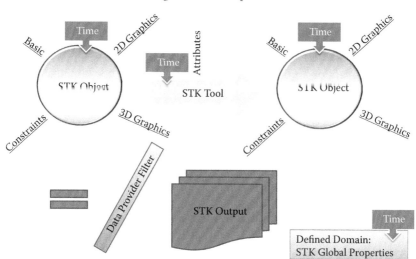

**FIGURE 7.2**
Full ontology.

This law of geography mirrors the laws of physics. For central bodies and applied physics, we could look at Newton's "Inverse Square Law," where amount or intensity of a dynamic effect is inversely proportional to the distance of the objects. The blend of sciences from the fields of physics, computer science modeling, and geographical application has given us the field of study called "spatial temporal analytics." Here, with spatial information systems, we tend to blend the lines of the science of geographical modeling and the application of physics via computer simulation. This unique blend between geographical modeling and physics applications helps us visually and analytically be able to observe physics-based dynamics in a pre-mission or post-mission analysis situation. When we include the temporal aspect (time), then we discover that relations become even more complex.

Spatial temporal analytics allows us to qualitatively and quantitatively evaluate the relationships of near and far items in their diversity and their unique likenesses. Time is an integral part of every scenario, object, and tool within the software. This is important to understand. Time intervals can change the dynamic of the analysis at any level, either globally or with any of the local objects or tools. With the element of time embedded both globally in the entire cartographic analysis and also locally in each object that is introduced within the analysis time frame, each element within the ontological study becomes temporally sensitive. By placing the focus on the fourth dimension, it moves to an object that is driven by the time-clock element that is central to STK. This gives strength to the STIS's capability of analyzing events not only spatially but also over distance, time, and orientation.

## Point, Point Mass, and Vector Geometry

When we use coordinate systems to represent a physical model, we often use a point. The point represents a specific coordinate position at a specific moment in time. For a position on land, we would use coordinates of longitude, latitude, and altitude. With this information, understanding the reference frame we are using and the position where we have set the point, we can find and use the location of the object the point is representing. In STK, we use a point to represent static objects such as a target or a facility.

When we add vector and force analysis to a point and give the point motion so it is changing its location over time, we consider this a point mass. The point mass represents the center of gravity of a moving vehicle object, such as a ground vehicle, ship, aircraft, satellite, launch vehicle, or missile. In theory, the mass point within the STK software represents the classical mechanics calculations of forces: drag, direction, velocity, and more. The fidelity and accuracy of the results of these calculations are user defined within the object's properties. It is within these properties that the user sets the equation's values and constraints for each object. This gives the user options to stay within the laws of physics or to simply use general types of equations because the amount of fidelity is appropriate for the given analysis. At times, using nondefined properties allows a user to completely defy the physical laws and place ground vehicles on the moon! It is important to define your parameters.

For instance, if I had a standard ground vehicle and I set the properties up for this vehicle to move toward the (+z) altitude of 10,000 kilometers at a velocity of 1,000 km/hr, I could do this with the default parameters using a Great Arc propagator and no constraints or parameters set. The default parameters do not keep my analysis from being unrealistic. This is fine if I just need to show the vehicle on a mountain top or pretending that the vehicle is in low Earth orbit for a positional reference. I could lower the altitude to constrain the vehicle to follow the terrain environment of the Earth. This would require the vehicle to interpolate the triangulated irregular network (TIN) frame of the terrain as correlating waypoints to represent the vehicle's changing position of longitude, latitude, or altitude over time. This information would be defined within the route propagator and the Terrain page of the object.

Vector geometry within the STK Object allows you to define motion. Often, STK handles analytics and visualization separately. With vector geometry, this is also true. The vector geometry is built into the propagators of the object by default. To visualize your vector geometry within the software, you need to toggle on the visualization parameters found within the 3D Graphics Vector page. If you want to create custom vectors, you may do this by using the Vector Geometry Tool (VGT). The VGT is a nonontological tool and is discussed later on in the Tools section. STK applies the sophistication of vector geometry added to the inherent properties that support definitions of position in time and space, as well as motion, velocity, and orientation. If you need a

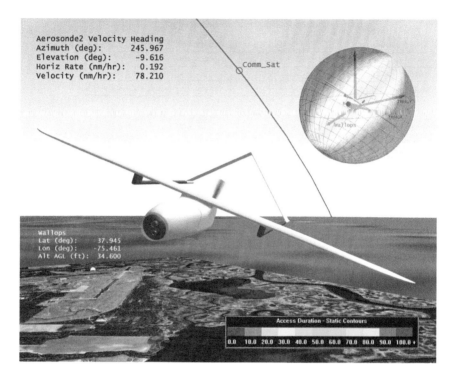

**FIGURE 7.3**
This animation is a simulation of two UAVs and a satellite in motion over Wallops Island, Virginia. We can run access and communication evaluations on all of these at once if we have refined our objects properties well enough.

more remedial and broader overview, a recommended primer on Newtonian physics can be found in Dr. Jerry Sellers's book, *Understanding Space*.

## Moving Vehicle Propagators

The strength behind STK software is the robust and well-defined attribute development of each object. This is why we use ontology to refine the object in the manner that best fits the need of analysis. With a moving object, the level of fidelity needed to represent the physics-based motion is based on the definition of the propagator. As you refine the properties for your moving vehicle, factors of velocity are evaluated, and even drag, lift, and thrust are considered in the more robust force models. The more we refine our properties, the more forces and physics terms we add to our computation for motion within the given environment. STK has the unique ability to apply given force models for the appropriate object propagator and the domain it appears in. This allows us

**Understanding Position and Time together**

| Time | Position A | Delta t | Position B |
|---|---|---|---|
| 06:00 | $x, y, z + x', y', x' + x'', y', z''$ | 10 | $x_1, y_1, z_1 + x_1', y_1', x_1' + x_1'', y_1'', z_1''$ |
| 06:10 | $x_1, y_1, z_1 + x_1', y_1', x_1' + x_1'', y_1'', z_1''$ | 10 | $x_2, y_2, z_2 + x_2', y_2', x_2' + x_2'', y_2'', z_2''$ |
| 06:20 | $x_2, y_2, z_2 + x_2', y_2', x_2' + x_2'', y_2'', z_2''$ | 10 | $x_3, y_3, z_3 + x_3', y_3', x_3' + x_3'', y_3'', z_3''$ |

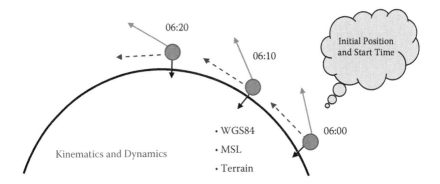

**FIGURE 7.4**
Understanding Time and Position.

to control the level of computational resources needed to compute versus the level of fidelity of computation required for the analysis (see Figure 7.4).

STK dynamics allows multiple moving and static objects to use multiple types of tools within multiple levels of environments (Earth, sky, as well as space) and calculates all of them over time during the same analysis interval. A typical way for handling this intensive form of computation is to use interpolation during selected time-step intervals and based on the selected granularity of the terrain or gravitational model. This is the strength of STK.

## Ground Vehicles

The Ground Vehicle object primarily propagates using the Great Arc (Great Circle algorithm). It is based on point-to-point computations. The Ground Vehicle object also uses ephemeris or live data feeds, as is often used for Command, Control, Communications, Computers, Intelligence, Surveillance, and Reconnaissance (C4ISR) modeling. The Ground Vehicle is the simplest of the Moving Objects category.

## Aircraft

The Aircraft object is robust and models arcs using Analytic Curves. The propagators range from the simple Great Arc to Aircraft Mission Modeler (AMM). As a point mass, the Aircraft object has Attitude properties for all propagators; however, if the AMM is engaged, then the default Attitude capabilities are overridden by the AMM features. There is more information about the Aircraft object in Chapter 8.

## Satellites

The Satellite object is defined with multiple propagators and has strong modelling capabilities. The Analytic, Semi-Analytic, and Numerically Integrated propagation choices not only give you the ability to define how accurate your analysis needs to be in comparison to the computational expense of your resources, but they also have specialized constraint capabilities that allow for geostationary belt exclusions, terrain grazing, and more. See Chapters 9 and 10, which dig deeper into the attributes of the Satellite object and how that affects the modeling behavior of this particular point mass.

## Launch Vehicles

The Launch Vehicle, as a point mass, uses low-level attribual properties to model the launching sequences of a vehicle from launch point to orbit insertion. The computation of an ascent trajectory uses propagators of SimpleAscent, STK External (ephemeris), or a real-time connection. The basic default properties to give a ballpark analysis have their own unique and robust specialized propagation tools. They all consider thrust, drag, lift, and gravity as well as yaw, pitch, and roll. They are all used for their designated purposes.

Launch Vehicle has unique capabilities of being able to allow other vehicles to "follow" the same flight path. The Launch Vehicle is used with the Astrogator or Missile Modeling Tool (MMT) add-on propagators for Satellites and Missiles respectively. For instance, if we were to recreate a historical space shuttle launch, we would use Launch Vehicle to model the solid rocket boosters and model the space shuttle as a heavy spacecraft. The space shuttle would use a "launch to follow sequence" setup with the correct propagator, Astrogator. The Launch Vehicle would need consider the extra gravitational pull from the mass of the shuttle and consider the loss of the mass at separation. Launch Vehicle goes beyond the scope of this book.

## Missiles

Missile objects also have unique propagation capabilities. Missile propagators include Ballistic, STK External, Two Body, High-Precision Orbit Propagator (HPOP), and RealTime data feeds. The Missile Modeling Tool (MMT) does not come standard with the STK software. However, using MMT with other tools within STK allows for sophisticated flight modeling capabilities. The ability to use end-to-end trajectory analysis with kinetic interceptor and integrated missile-defense design gives accurate missile modeling scenarios. As an air vehicle, the Missile object handles the forces

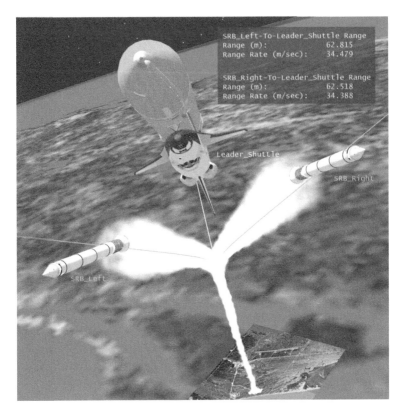

**FIGURE 7.5**
The Launch to Follow Sequence is used withing Astrogator in conjunction with objects, includ-ing rockets and missiles.

of flight. It is capable of trajectory simulations and phase analysis for the path. MMT leverages robust features that go beyond the scope of this book.

## MTO

Both moving and static objects relate to each other dynamically when inter-visibility tools are used. If you use Multi-Tracking Objects (MTOs) to rep-resent your STK objects, the number of visualized and analyzed objects is massive. Full intelligence, surveillance, and reconnaissance (ISR) types of analysis can be utilized with STK.

If you would like to learn more about vehicle propagation and how Newton's laws are calculated for *N*-Body motion within space, I would recommend reading David Vallado's book, *Fundamentals of Astrodynamics and Applications*.

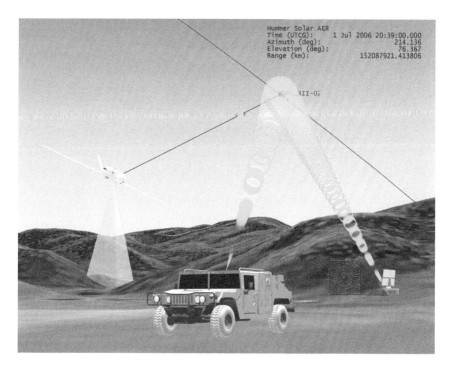

Hummer Solar AER
Time (UTCG):      1 Jul 2006 20:39:00.000
Azimuth (deg):                    214.136
Elevation (deg):                   76.367
Range (km):             152087921.413806

**FIGURE 7.6**
Static and moving objects create dynamic relationships.

# 8

## Aircraft

### Objectives of This Chapter

- The Aircraft Object
- Aircraft Mission Modeler
- Launch Vehicles (Limited)

### Aircraft Object Overview

The STK Aircraft object models the flight of an aircraft using the simplistic Great Arc propagator as well as the sophisticated Aircraft Mission Modeler. As a point mass, it uses the center of gravity of an object; the air vehicles also model types of aircraft and their unique physics parameters. With a Spatial Temporal Information System like STK, we can evaluate the effects of variable geometry of flight characteristics on the point mass. We all know that a full-scale B-52 handles differently than an unmanned aerial vehicle (UAV) or even a helicopter. STK can use enhanced performance models within the mission modeler of the software to represent flight paths and flight performance.

STK air vehicles consist of the Aircraft, Launch Vehicle, and Missile objects. Generically, they all model flight from the perspective of dynamics that STK air objects model. Air objects model positions and orientation of assets. They reference the Earth terrain and gravity model for flight behavior. Because of the ability to model flight and air vehicle position and orientation, you can evaluate your mission in regard to your payload, your frequency, and terrain. You are able to evaluate sensor coverage areas, signal strength, and terrain when communication child objects are used.

With more than one air vehicle in the sky, you can evaluate the relationships between vehicles and their routes. An example of this form of geometric analysis is the way STK handles an aircraft's ability to deconflict routes by computing an air vehicle's route, horizontal and altitude separation regarding proximity, and basic trajectory prediction. Since all aircraft performance models have unique physics-based parameters and propagation capabilities, it is important to understand the attribute details to create a realistic evaluation. For this chapter, we further explore the Aircraft object and introduce Launch Vehicles. Essentially, once you learn the basics of air vehicles, it is easy to transfer that knowledge to the other vehicles.

## The Aircraft Object

Your default aircraft object uses physics-based vector geometry properties within the graphic representation of point mass geometry and a polyline to visualize the position and propagation of the object. The basic default route propagation capabilities use the Great Arc propagator and a generic point mass object with a mesh model for visualization. However, this is easily modified and becomes more robust when you select the Mission Modeler route propagator and then begin to refine attributes to the aircraft performance models (see Figure 8.1).

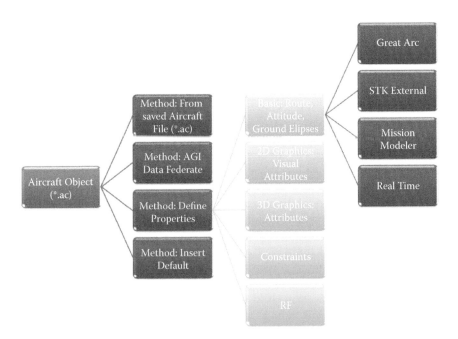

**FIGURE 8.1**
*.ac object hierarchy.

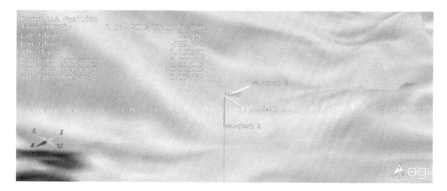

**FIGURE 8.2**
The point mass STK Aircraft Object has default vector geometry to use during propagation computation to determine the position of the object.

**FIGURE 8.3**
Raven with body.

To create an Aircraft object, there are four primary methods to choose from: a saved Aircraft file, the AGI Data Federate library, the Aircraft Properties option, or inserting a default aircraft using the Great Arc propagator. With all of these methods, you may modify the aircraft properties after it is brought into the object browser with STK.

## Aircraft Route Propagators

Within the Basic Route properties page, the propagators determine the degree of accuracy and fidelity the aircraft or UAV flight represents. The Aircraft object has several important basic properties that include the route for setting the propagators, attitude profile definition, and the ability to add

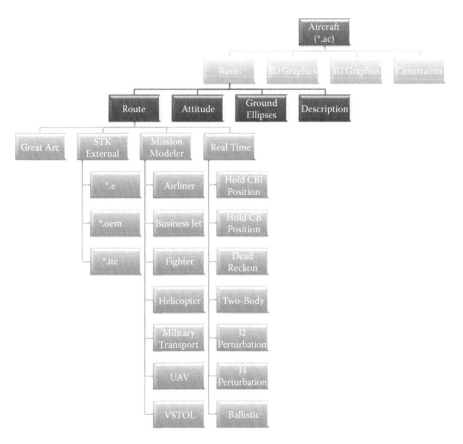

**FIGURE 8.4**
The aircraft route propagators.

a set of ground ellipses. The route properties can be set using one of four propagator types: Great Arc, STKExternal, Mission Modeler, and Real Time (see Figure 8.4).

For simplified analysis, the Great Arc propagator (default) reflects latitude, longitude, and altitude (*x, y, z*) positioning of the craft as it moves from one waypoint to another, denoting speed, acceleration, and turn radius at a minimum. STKExternal allows the user to import ephemeris files that directly empower the Aircraft object. These data feeds may be in STK and end in a *.e extension or in Consultative Committee for Space Data Systems (CCSDS) and end in a *.oem extension. USGOV licensees may also import *.itc ephemeris files if needed.

If you are not predicting route propagation but are using ephemeris for positional data, STK Aircraft uses ephemeris feeds well. Because most ephemerides have the same basic data for multiple types of air vehicles, it is probably a good idea to make sure you are using the correct type of vehicular data for the object you are modeling, i.e., an aircraft.e file for an Aircraft object. If you have the incorrect vehicle ephemeris, you might garner unexpected results from your vehicular path.

Additionally, the Real Time propagators allow for data feeds to be brought in with telemetry, Look Ahead data, or data that can calculate projected positions using interpolation. All data feeds may be archived. The Look Ahead properties use several options; Real Time propagation uses STK Connect to hook into the software. When you use the near Real Time propagator, it is considered a best practice to modify the Global properties within the scenario level on the Basic/Time page and set it to Mean Sea Level for a better flight performance.

To customize your flight to model a closer representation to true-to-life flight with banking, and following the terrain, vertical, and short takeoff and landing styles with specified route procedures, we use the Aircraft Mission Modeler (AMM). This gives you the features needed to recreate a realistic flight. AMM models aircraft maneuverability to understand solid pre- and post-mission analysis. Figure 8.5 and Figure 8.6 represent two sample flights from AGI scenario examples to show some of the powerful route sequencing.

STK evaluates route prediction using Analytic Curves. From the point-to-point segments, the positional values are generated and stored in a flat-Earth coordinate system. This method is not as computationally expensive and cumbersome as true flight simulations are. This is used to lessen the burden on the computer processor during route prediction and flight performance based on defined aircraft parameters, but still have enough memory to effec-

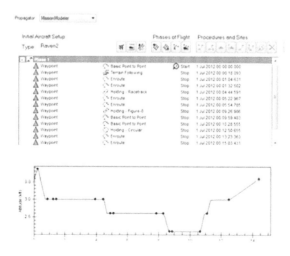

**FIGURE 8.5**
AMM Flight 1.

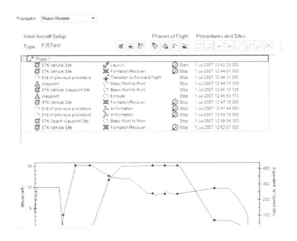

**FIGURE 8.6**
AMM Flight 2.

tively compute the sensor and communications activity. Arcs are sufficiently represented in the flight path without the intensive burden on memory.

## Aircraft Mission Modeler

The Mission Modeler propagation has become robust and powerful. The aircraft performance, when the correct vehicle is matched to the correct object, has been proven accurate on many levels. Sensis Corporation, now known as Saab Sensis Corporation, provides airline and airport operations management support and has evaluated the Aircraft Mission Modeler for verification and validation (V&V). The V&V used a Boeing 747-200 series aircraft model and compared results to the known performance characteristics: airspeed, altitude versus true airspeed, altitude versus time, and mission time against the procedures from the AMM models during takeoff, climb, acceleration, cruise, descent, and landing.

Their report stated, "The trajectory associated with the B-747 model developed in STK is close to the external 'gold' trajectory with some differences" (Project Summary, *STK B-747 Aircraft,* 2005). Copies of this detailed V&V report are available from vv@agi.com. Since this report has been published, significant improvements and refinement to the AMM product have been made, primarily with version 9.0.

Aircraft Mission Modeler attitude is determined from expert systems logic. This software module is designed to be set up the same way a fighter pilot thinks about the mission. Tom Neely, aerospace engineer and retired

U.S. Air Force F-14 pilot, is one of AGI's subject matter experts when it comes to AMM. He describes AMM as

> The [integration of] relationships between Mission, Phase, Procedure, Site, Aircraft, Performance models. In reality, everything is a plugin. AMM is a large expert system wrapped around a sophisticated flight mechanics engine. Procedures determine the behavior of the aircraft in conjunction with performance models that define the constraints on the capabilities of the aircraft. Procedures are defined relative to a generic Site object which are often a waypoint or a runway but could be anything. An aircraft is a collection of performance models of various types. The aircraft is designed to model combat performance, max range/endurance cruise performance, anything in between. Procedures are contained in phases where the user must select the single performance model of a given type. Each type must specify a single way to takeoff, climb, cruise, descend, land, etc. This allows the typical combat missions to be modeled "as fly to station," "engage in combat," "return to base," etc.

AMM has a catalog of several aviation performance models to select from: Airliner, Business Jet, Fighter, General Aviation, Helicopter, Military Transport, Turboprop, Unmanned Aerial Vehicle (UAV), or a Vertical or Short Take Off or Landing vehicle (VSTOL). Each model has differing physics capabilities within the Performance Models properties page built within each model to allow more accurate flight modeling and simulation. Performance models are customizable to match the unique properties of your particular aircraft. As each Performance Model page is refined to match your spe-

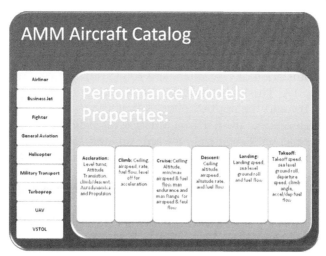

**FIGURE 8.7**
AMM performance models.

**FIGURE 8.8**
AMM toolbar.

cific performance capabilities, you may select Help at any time within your software to understand how STK computes these properties within AMM.

Aircraft behavior handles best when the flight dynamics are computed within the Mean Sea Level (MSL) surface reference for the Earth's globe (see Figure 8.7). The global settings are located in the root of the scenario in the 3D Graphic Global Attributes page. While the geoid algorithm using the default WGS84 ellipsoid does work, the calculations using the MSL as the surface reference give a more realistic calculation as the positional changes in the aircraft are interpolated.

AMM Route Development is created by using the Phases and Procedures of Flight based on the initial state of the aircraft found within the properties page or it may be created and modified by using the 3D Mission Editing tool (see Figure 8.8) within the 3D Graphic window. Additionally, Digital Aeronautical Flight Information Files (DAFIF) are easily used within STK for AMM route development using either the aircraft Basic/Route properties page or the 3D Mission Editor tool. DAFIF data may also be used within STK for visualization within the 2D and 3D Graphic window properties. In addition, the files may be used with the GIS Analyst and Analysis STK modules. DAFIF files are currently only available through the National Geospatial-Intelligence Agency (NGA).

## Vehicle Translation in AMM

The motion translation of yaw, pitch, and roll (YPR) is important to understand within STK. This is enabled by the combination of the point mass vector geometry and the applied physics that allows modeling of aircraft behavior. In reference to the Earth, evaluation regarding the position of the object is essential. YPR includes forward, up, and left, and then rotation motion is modeled with extensive computing. STK evaluates yaw, pitch, and roll without the use of standard Euler angles but still using the concept of a three-rotation sequence. Here (see Figure 8.9), the method disregards which axes are moving but runs the YPR sequence from the axes of $X'$, $Y'$, and $Z'$. Yaw is the rotation about the $Z'$ axis, whereas the direction of motion would be $ZXY$, $ZXZ$, $ZYX$, $ZYZ$. Pitch refers to the rotation about the $Y'$ axis, whereas the direction of motion would be $YXY$, $YXZ$, $YZX$, $YZY$. Roll is the rotation around the $X'$ axis, whereas the direction of motion would be $XYX$, $XYZ$, $XZX$, $XZY$. All axis rotation is considered

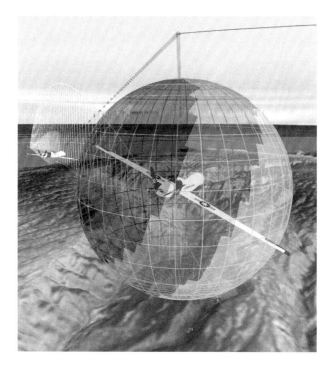

**FIGURE 8.9**
Aircraft evaluating access and communications.

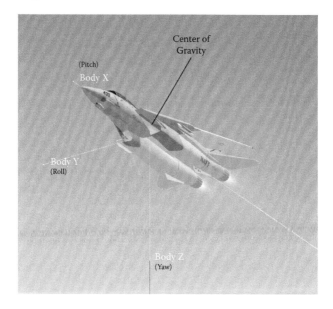

**FIGURE 8.10**
YPR: Axis of rotation followed by direction of motion.

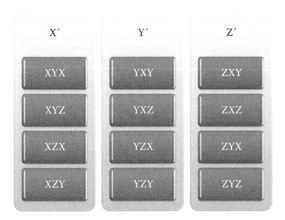

**FIGURE 8.11**
YPR rotation table (counterclockwise).

counterclockwise (see Figure 8.11). Essentially, it represents the angle and the vector as it defines the rotations around the axis using the right-hand rule. An advanced explanation of the sequence rotations and the way AMM handles motion is defined within the Help files.

# 9

## Satellites

### Objectives of This Chapter

- Definition of a Satellite
- STK Satellite Object
- Satellite Propagator
- Data Sources Affecting the Selection of a Propagator

### What Is a Satellite?

In STK, a satellite is a moving vehicle object that moves around another object, primarily under the influence of gravity. For example, the International Space Station (ISS) is a satellite that orbits the Earth. While other forces—such as atmospheric drag or solar radiation pressure—can affect the motion of a satellite, these are considered to be minor forces compared to the main gravitational force of the central body, and they are known as perturbations.

### Creating a Satellite in STK

The Satellite object has many methods available to insert it into a scenario. However, before using any of these methods, it is very important to make sure that STK has the most current data available. AGI regularly updates satellite databases, GPS almanacs, Earth orientation parameters, and space weather data throughout each day and all of that data is made available to STK users via the Data Update Utility, which can be accessed under the Utilities pull-down menu item. The Data Update Utility affects the scenario

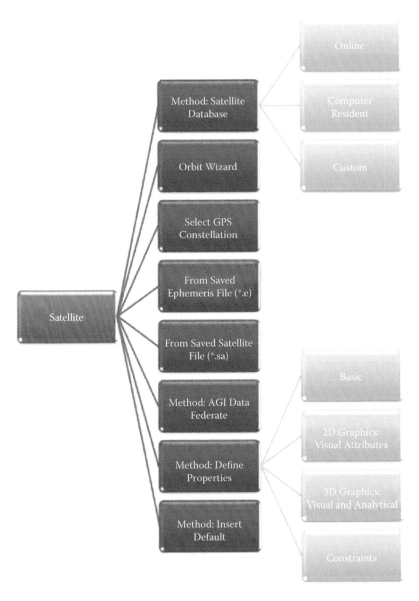

**FIGURE 9.1**
Satellite hierarchy.

and the satellite information globally in the scenario. Other modifications are made locally to each satellite as needed.

The Data Update Utility can be run manually or set up to run automatically. When run manually, any local data that has more current data available on the AGI system is shown in red. Simply check the files you wish to download and click "Update Now." To set up automatic updates (highly

**FIGURE 9.2**
Satellite data update utility.

recommended), simply check "Enable Automatic Updates," select the files to update, and specify the frequency and a start time. Then click OK or Apply. That's all there is to it!

## Insert Satellite from Database

Once we have the latest data, perhaps the easiest way to add a satellite is to use the Select from Satellite Database option found under the first two

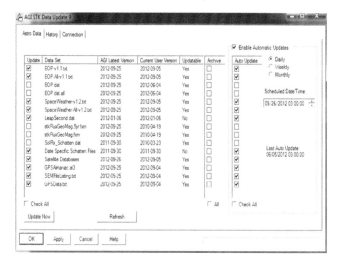

**FIGURE 9.3**
Update utility before update.

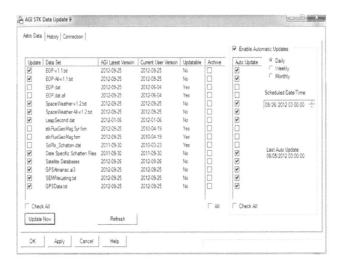

**FIGURE 9.4**
Update utility after update.

methods for creating an object found when you select the satellite object. The "Select from Database" approach uses NORAD two-line element (TLE) sets and the SGP4 orbital propagator to provide medium-fidelity force modeling. The reason this approach is so convenient is that there are data available for almost everything tracked by the US Space Surveillance Network (SSN). As of this time, that's over 15,000 objects in Earth orbit.

The Insert from Satellite Database method brings up a dialog box that allows you to search for satellites using their name (or portions thereof), an SSC number, or advanced search criteria. You can also specify to get the data from the AGI Server to ensure you get the latest TLE data or specify to use other options, such as loading from a file on your local computer.

Note that one of the default search criteria is for the satellite to be active, so don't forget to uncheck this using the Advanced Filter if you are looking for dead satellites, rocket bodies, or debris. Also, if no TLE data is available for a particular object, a search result will not be returned unless you change the configuration using the "Modify" button and uncheck "Filter search results by TLE availability." This item is checked by default to avoid getting an error message when trying to add a satellite with no TLE data.

## GPS Satellites

If you are adding one or more GPS satellites to a scenario, a better way to do that is to insert the satellites from a GPS almanac. Almanacs are special orbital data files used by GPS receivers to determine the orbital positions of all the operational satellites in the GPS constellation for use in obtaining an initial navigational fix. The propagator only uses a basic two-body Keplerian

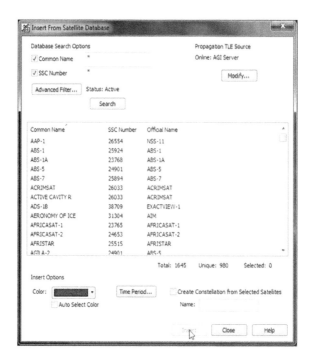

**FIGURE 9.5**
Insert from satellite database.

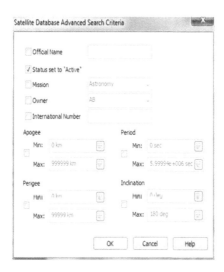

**FIGURE 9.6**
Satellite advance search criteria.

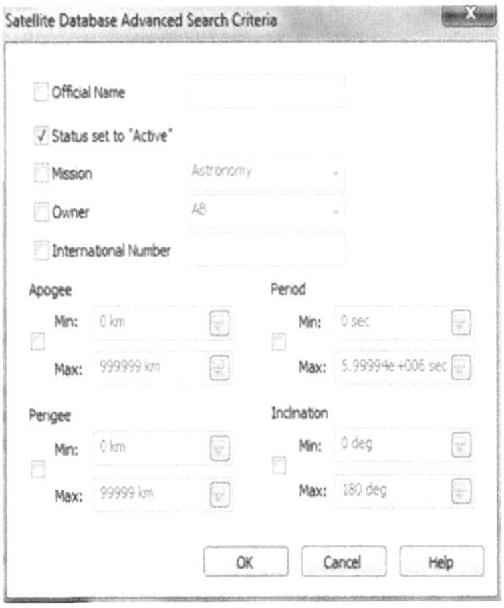

**FIGURE 9.7**
Satellite database TLE source.

force model, but the GPS orbital regime permits this simplification. This results in orbital positions about an order of magnitude more accurate than those available using TLEs. Selecting "Load GPS Constellation" makes it quite easy to load the entire GPS constellation at once, using the latest almanac data. If you are unfamiliar with the STK Constellation object, please refer to Chapter 12 on constellations.

Once the "Insert from GPS Almanac" panel is up, simply specify the source of the data and select the satellites needed. Each satellite is identified by the PRN, SVN, and SSC numbers.

## Orbit Wizard

The Orbit wizard can be found in the Insert STK Objects option under the Satellite pull-down menu item. This tool provides a slick visual interface to define a variety of standard orbit types (Circular; Critically Inclined; Critically Inclined, Sun Sync; Geosynchronous; Molniya; Orbit Designer; Repeating Ground Trace; Repeating Sun Sync; and Sun Synchronous). The Orbit Designer type allows you to adjust each of the individual Keplerian elements—either by manually entering a value or using the slider bars—as can be seen in Figure 9.9. Using the slider bars is a great way to see the effect of changing these orbital parameters directly on the ground trace. The other types fix the parameters specific to that type (e.g., eccentricity equal to zero for the Circular type) while allowing you to vary the rest. The Orbit wizard provides an easy way to create many basic orbit types and is a wonderful educational tool as well.

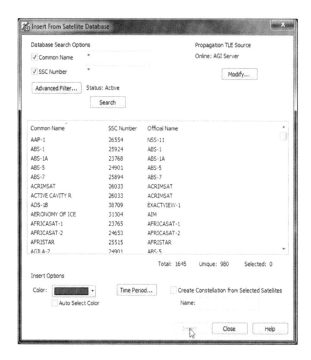

**FIGURE 9.8**
Insert from a satellite database.

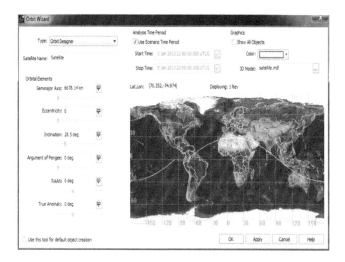

**FIGURE 9.9**
Orbit wizard.

**Insert from Saved External Ephemeris File**

The "Insert from Saved External Ephemeris File" option allows you to load in a satellite ephemeris file—a file containing time-stamped Cartesian or spherical (e.g., latitude, longitude, and altitude) state vectors—using an STK-defined format. This format, which is defined in detail in the STK Help files, can be used to model orbits using virtually any force model, including maneuvers. In fact, it is easy to take virtually any text ephemeris file, tweak the format using a text editor, and then provide a couple of lines of metadata to describe the coordinate frame and time system to input data from virtually any force model and coordinate system. Load that ephemeris into STK and it makes sure all orbits are represented in a consistent coordinate frame for analysis. This approach was used in the early implementations of the Space Data Center to ensure all satellites were represented in a common coordinate frame and that all maneuvers—including ones generated using ion thrusters—were incorporated accurately.

**Satellite Objects**

You can always simply start by creating a Satellite object and then setting the orbital parameters via the properties panel. Each type of propagator available in STK can be selected on the Orbit tab via a pull-down box. The parameters and options for many propagators are the same or similar.

For example, TwoBody, J2Perturbation, J4Perturbation, HPOP, and LOP each require defining the Coordinate Type (Classical, Equinoctial, Delaunay Variables, Mixed Spherical, and Spherical) and System (Fixed, ICRF, MeanOfDate, MeanOfEpoch, TrueOfDate, TrueOfEpoch, B1950, TEMEOfEpoch, TEMEOfDate, AlignmentAtEpoch, and J2000), setting the epoch, and then setting the individual state values. There are plenty of options (and flexibility), and the details for each option can be found in the extensive STK Help files.

For the SGP4, GPS, STKExternal, and SPICE propagators, files in a specific format must be loaded and, in the first two cases, the data can be obtained directly from AGI's servers. The SPICE format allows modeling of complex deep-space missions (e.g., Lunar Reconnaissance Orbiter) using the data from sources such as NASA's Planetary Data System (http://naif.jpl.nasa.gov/naif/data_archived.html).

Astrogator—a very powerful but specialized tool for interactive maneuver and trajectory design—requires its own chapter (Chapter 10).

**Satellite Propagation**

To define the motion of a satellite—or to propagate it—you must choose an appropriate force model to represent the combination of forces acting on the satellite. STK provides a broad range of propagators that define a satellite's motion under the influence of a specific force model. These range from a simple Keplerian two-body force model all the way up to ones that include complex nonuniform central gravity fields; third-body effects from the sun, moon, and planets; atmospheric drag; solar radiation pressure; and even the motion of the central body's crust using the High-Precision Orbit Propagator (HPOP). Users may also model other forces (such as ion propulsion) with custom software and include that motion as an external ephemeris using the STKExternal or SPICE ("The SPICE Concept," http://naif.jpl.nasa.gov /naif/ spicecon cept.html) formats.

The selection of an appropriate propagator depends on the objectives of the analysis. To demonstrate the basic motion of satellites with different orbital properties, such as inclination or eccentricity, a Keplerian two-body model is probably sufficient. To illustrate the secular effects—or slow non-periodic changes—due to the Earth's oblate spheroidal shape would require a J2 propagator. And to predict when a satellite in low Earth orbit will be visible for observing from the ground requires a model that incorporates the effects of atmospheric drag. In each case, the user must balance the required precision of the results against the computational burden required for higher fidelity force models.

## Classes of Propagators

In general, propagators fall in three basic categories: analytic, semianalytic, and numerical. Numerical propagators numerically integrate all of the forces acting on a satellite and can be used with very detailed force models and allow great flexibility in the choice of the forces modeled. However, numerical propagators are more computationally intensive and must integrate from some specific initial state to determine the state at any other time. STK provides two types of numerical propagators: HPOP and Astrogator (which is discussed in Chapter 10).

Analytic propagators reduce the computational burden by developing a model that allows you to directly determine the state at any point in time without the need to numerically integrate the forces. To develop these analytic models requires limiting the forces modeled and applying assumptions about the types of orbits (e.g., low eccentricity). As a result, while analytic propagators are faster, they tend to be less accurate and provide less (or no) flexibility in the choice of forces modeled. Examples in STK are the two-body Keplerian, the J2 and J4 Perturbations, the SGP4 (specifically, the near-Earth portion, with orbital periods less than 225 minutes), and GPS almanac propagators.

Semianalytic propagators combine the speed of analytic propagators for most of the force model with numerical integration of those forces that don't easily permit analytic development. A notable example is the use of SGP4 (or more accurately, SDP4) for orbits with periods greater than 225 minutes. STK also provides the Long-Term Orbit Propagator (LOP) and Lifetime, both of which combine analytic and numerical propagation.

## Accuracy of Propagators

The fidelity of the force model for a given propagator determines how accurately a satellite's position can be propagated away from some initial condition. To give some idea of the magnitude of these differences, compare the results from the Two Body, J2 Perturbations, J4 Perturbations, SGP4 (fit using two days of the truth orbit), and HPOP propagators for an assumed known initial condition for a number of orbit types.

Comparative analysis assumptions:

- Orbit types: LEO1 (circular 420-km altitude, 51.6-degree inclination), LEO2 (circular 840-km altitude, 98.6-degree inclination), MEO (circular 20,200-km altitude, 55-degree inclination), GEO (geostationary orbit)
- Sphere, 1 m diameter, 10 kg, coefficient of drag 2.2
- Only gravitational potential and drag (using Jacchia 1970) are modeled
- Truth: HPOP with EGM96 $70 \times 70$ gravitational field, solar/lunar gravitational perturbations, and Jacchia 1970 drag

The results shown in the table are interesting, if not in some cases counterintuitive. For the LEO1 case, the differences grow until they reach a maximum after seven days of propagation (positions are on opposite sides of the Earth). As might be expected, the two-body result is worse than the J2 and J4 perturbations.

For the LEO2 case, however, the two-body effect is actually better, in part because of the secular effect of J2 on the orbit's line of apsides (the line between the ascending and descending nodes). Overall, all three of these propagators do better because less drag is experienced in higher orbit.

For the MEO case, the differences are much smaller, with the J2 and J4 results only slightly better than the two-body results, since this orbit experiences no drag and the J2 and J4 perturbations are significantly reduced with altitude.

Finally, in the GEO case, the differences are the smallest, with the relative differences in line with what we might expect. We could, however, see much different results if the longitude (not right ascension) of the ascending node were near one of the unstable equilibrium points in Earth's gravitational field.

Overall, probably the most counterintuitive result of this rudimentary analysis is that the SGP4 results do not differ significantly from the HPOP

**TABLE 9.1**

LEO, MEO, and GEO Comparative Table

| Days | LEO1 | | | | LEO2 | | | | MEO | | | | GEO | | | |
|---|---|---|---|---|---|---|---|---|---|---|---|---|---|---|---|---|
| | 2-Body | J2 | J4 | SGP4 | 2-Body | J2 | J4 | SGP4 | 2-Body | J2 | J4 | SGP4 | 2-Body | J2 | J4 | SGP4 |
| 0 | 0 | 0 | 0 | 0 | 0 | 0 | 0 | 0 | 0 | 0 | 0 | 0 | 0 | ■ | 0 | 1 |
| 1 | 1309 | 1078 | 1079 | 1 | 434 | 1187 | 1187 | 0 | 35 | 32 | 32 | 0 | 12 | 8 | 8 | 1 |
| 2 | 2986 | 2559 | 2561 | 3 | 870 | 2371 | 2371 | 0 | 70 | 63 | 63 | 0 | 24 | 15 | 15 | 1 |
| 3 | 5010 | 4437 | 4440 | 16 | 1357 | 3550 | 3550 | 1 | 104 | 95 | 95 | 0 | 36 | 25 | 24 | 2 |
| 4 | 7260 | 6618 | 6622 | 42 | 1777 | 4706 | 4705 | 0 | 138 | 126 | 126 | 1 | 47 | 32 | 32 | 5 |
| 5 | 9569 | 8977 | 8981 | 84 | 2199 | 5823 | 5822 | 1 | 172 | 157 | 157 | 1 | 59 | 40 | 40 | 8 |
| 6 | 11599 | 11215 | 11219 | 163 | 2709 | 6895 | 6893 | 2 | 206 | 188 | 188 | 1 | 71 | 47 | 47 | 12 |
| 7 | 12930 | 12936 | 12938 | 267 | 3091 | 7918 | 7916 | 3 | 240 | 220 | 220 | 2 | 85 | 54 | 54 | 16 |

*Note:* This chart explains the fidelity of the force model for a given propagator determines how accurately a satellite's position can be propagated away from some initial condition.

results except for the LEO1 case. The LEO1 difference is due primarily to the very limited drag model in SGP4. But these comparisons are between force models and not between a force model and reality. Choosing a different drag model for HPOP would yield different results.

In practical usage, though, the quality of an SGP4 orbit for the other cases is more the result of limitations in the observations used to produce the orbit (the density and accuracy) than in the model itself. In this analysis, we fit observations directly from the truth model and achieved very good agreement. The lesson here is that even the best force model is limited by the quality of the observations used with it.

A much more in-depth analysis of the differences of a variety of force models can be found in David Vallado's "An Analysis of State Vector Prediction Accuracy" (available at http://www.centerforspace.com/downloads/files/pubs/USR-07-S6.1.pdf).

## How Data Availability Affects the Choice of a Propagator

Selection of a propagator also depends on the availability of data needed by the propagator. It does little good to select HPOP to model the motion of a satellite if you don't have any high-precision data to start with. In fact, STK allows the user to automatically select the appropriate propagator by simply choosing the types of data to be used. Not only does this make it easy to select an appropriate propagator, it also ensures that the correct propagator is used for those specific types of data, since each propagator is a model of reality that makes certain assumptions to reduce computational complexity. It is important to ensure that the data, which usually reflect the assumptions of the model, are then applied with the same set of assumptions used to create them.

## Configuring Other Satellite Attributes

Once the Satellite object has been created, there are still many properties that can be set to support detailed analyses. While a large number of these properties are common to other STK Vehicle objects, others are unique to the Satellite object and are examined more closely.

### Attitude

Although the general Attitude properties for the Satellite object are the same as for other Vehicles, there are many more standard types for Satellites. It is very important to ensure you understand the specific attitude alignment for your satellite if you need to analyze sensor pointing or solar panel tracking, or even if you are trying to calculate when glints occur off of specific surfaces (such as with iridium flares). The good news is that each of the available types is described in detail in the STK Help files.

## Eclipse Bodies

The time when a satellite is illuminated by the sun is often an important consideration, whether when trying to determine when a satellite is visible to an observer or when it will have to run off of batteries due to being affected by an eclipse. Although the primary source of an eclipse is the central body (e.g., the Earth for an Earth-orbiting satellite), other bodies can eclipse the satellite too. You can easily select those eclipsing bodies here for use in your analysis.

## Pass Break

The Pass Break property allows you to define where the beginning of each pass occurs. For example, the standard definition for the US SSN is that each pass (or rev) begins when the satellite crosses the Earth's equator heading north. The portion of the orbit from launch until the first ascending node is considered Rev 0, and Rev 1 begins when the satellite crosses the equator.

STK allows flexibility in deciding whether this boundary occurs at the equator (default) or any latitude or longitude, and STK can even select descending nodes for latitude crossings. It also allows you to specify the reference frame for the equatorial coordinate system (i.e., Inertial, Fixed, True of Date, or True of Orbit Epoch).

The Pass Numbering option permits flexibility as well, such as setting a specific pass number at a specific time or using the pass number in the data (e.g., from a TLE).

## Pass

The Pass Break definition is also used in the Pass property to determine which parts of an orbit are shown in the 2D or 3D windows. This action is accomplished primarily by setting the Lead and Trail portions of the Ground Track or Orbit Track. By default, STK will show one pass ahead and behind the satellite. The leading portion goes to the end of the pass and the trailing portion starts at the beginning of the pass (covering a full orbit).

Other options include All (the entire interval the object is defined for), Full (actually half a rev), Half (of Full or one-quarter rev), Quarter (of Full or one-eighth rev), Percent (of Full but less than 100 percent), or Time (which can be set to values smaller or larger than Full). Current Interval is the same as All, unless using Access or Custom Intervals (set under the 2D Graphics attributes), and then only the current interval is displayed. Note that it is easier to see the effects of each change by setting the Trailing Type to None and only changing the Lead Type.

No orbit is shown beyond the limits it is defined for, which is usually the same as the scenario Analysis Period. Those defined with external ephemerides (e.g., STKExternal or SPICE) may be defined for longer or shorter periods, however. In those cases, All or Current Interval will show the entire ephemeris.

Finally, the 3D Graphics Pass property has options to set Tick Marks on both the Ground and Orbit Tracks. The size of these marks is set by distance, while the interval is set by time.

### Orbit System

This property makes it easy to visualize an orbit in different reference frames. By default, STK sets this to Inertial by Window, showing the orbit in inertial space. However, Fixed by Window is quite useful for visualizing the orbits of geostationary satellites. An inertial representation of a geostationary orbit would look the same for all such satellites, showing the orbit going around the Earth's equator. But in a fixed representation, you can see how the orbit appears relative to a coordinate system rotating with the Earth, so the geostationary orbit looks like a dot (well, in reality usually a very small ellipse).

Other useful visualizations can be seen by selecting "Add VVLH System" to represent the orbit of one satellite relative to a second satellite. This option can be useful to show the relative motion of two satellites that are formation flying or in similar orbits. It can also be useful to compare two different

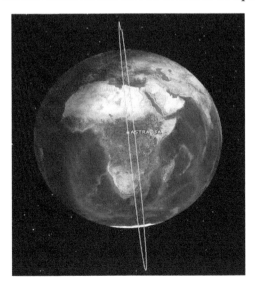

**FIGURE 9.10**
Earth-fixed geostationary orbit.

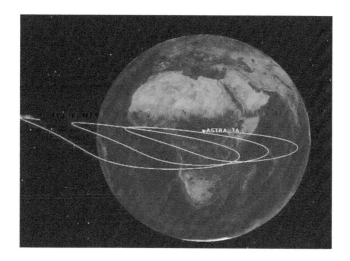

**FIGURE 9.11**
TLE orbit relative to ephemeris.

types of orbital data for the same satellite. The "Add VGT System" option allows you to create virtually any coordinate frame using STK's very powerful Vector Geometry Tool.

The Satellite object (see Figure 9.11) has an increased complexity based on the propagator chosen. As you continue to refine propagator styles, you might also consider looking at the Astrogator propagator. Astrogator is a separate module in STK and requires a special license. Astrogator is covered in a high-level manner in the next chapter.

# 10

## Advanced Satellites

### Astrogator

Astrogator is an add-on advanced propagator. From an ontological perspective, Astrogator refines the low-level attributes of the Satellite object. The module builds real-world mission design capabilities for the object. With this, you can maneuver your satellite to model rendezvous planning and space-based intercepts. Trajectory design handles libration points and orbit maintenance. It uses HPOP numerical integration as the fundamental orbit propagator. The module hooks through the Satellite/Orbit/Propagation path within the Satellite object class (see Figure 10.1).

Historically, Astrogator was originally designed as Swingby in 1989 and written by Computer Sciences Corporation as the PC replacement for the mainframe software Goddard Mission Analysis System (GMAS). This software was created for trajectory design and operations by the NASA Goddard ISEE-3/ICE mission. After many successful space programs using the software, Swingby was commercialized, branded, and sold as Navigator software. In 1996, Analytical Graphics purchased all rights to Swingby and Navigator. AGI integrated the software into their STK product line under the name Astrogator. The term "astrogator" was coined by science fiction writer Robert A. Heinlein in his book *Starman Jones*, written in 1953.

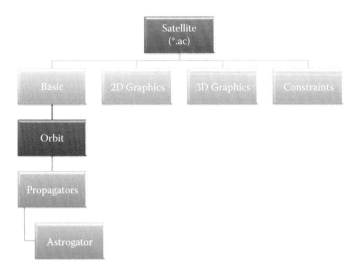

**FIGURE 10.1**
Hierarchy flow of satellite.

## Astrogator

Astrogator is used in the development of space mission planning and orbit modeling. It is accessed by opening the satellite propagators from the Basic/ Orbit page. With Astrogator, you can model satellite trajectory design, maneuvers, orbit transfers, and launch window analysis using visual programming language (VPL) techniques. VPL allows you to create commands using graphics as command segments that will function within a system. As the VPL code runs within the Mission Control Sequence, it also generates orbit ephemeris. Jonathan Lowe, aerospace engineer and systems engineer at Analytical Graphics, often describes Astrogator as "the modeling of a chapter in the life of a satellite." This is a very apt description of the nested programming features and functionality of the software module.

The Astrogator propagator, as a VPL, graphically introduces the spacecraft from an initial state. The Mission Control Sequence (MCS) toolbar, MCS windows, and the dynamic MCS attributes give the mission control analyst capabilities in design and optimized orbital mechanics maneuvers (see Figure 10.2). There are a variety of control segments and they are configured as:

- Initial State—as shown in Figure 10.3, details the initial conditions of the spacecraft; produces ephemeris
- Launch—simulates launching mode; produces ephemeris

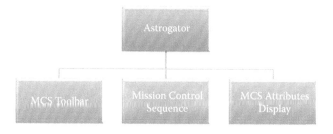

**FIGURE 10.2**
Astrogator.

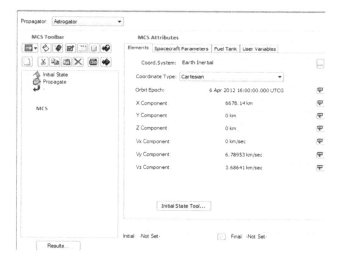

**FIGURE 10.3**
The Astrogator interface gives you the ability to control the Mission Control Segments and Sequences in a robust environment and by using Visual Programming Language.

- Follow—follows a leader vehicle until event enters; produces ephemeris
- Maneuver—impulse or finite; produces ephemeris
- Propagate—numerically integrates kinematic and dynamic motion until event enters; produces ephemeris
- Hold—holds position; produces ephemeris
- Run—run sequences forward or backward
- Target a sequence—defines the Differential Corrector, Design Explorer, Plugin Profiles, Scripting Tool Segment, and Segment Configuration
- Update—updates spacecraft attributes; produces ephemeris
- Return—returns to another position in the program
- Stop—stops all commands and analysis; end of program or nest

These series of segmented commands define the satellite's "life events," chapter by chapter.

The segments either produce ephemeris or trigger another event by allowing the final event results of the spacecraft to flow to the next segment as a newly defined initial state. Conditional stopping events trigger the behavior of the segment. For instance, you may have a satellite that will need to propagate until one of two conditions is met: a defined apoapsis or duration. As this satellite reaches the first served condition, either the apoapsis or the duration, it will create a final state for this segment and move the state to the next event where it will become the initial state of the satellite for the following segment. In other words, initiating the Run command compiles the Astrogator segments. This allows the Mission Control Sequence to systematically perform the segment functions defined by their constraints, attributes, and conditions.

## MCS Toolbar

The MCS Toolbar buttons allow you to build and run the Mission Control Sequences. The Help files define all the items on the page; this user reference guide will help you understand what is happening within the toolbar and give you the default values.

### Run Button

The Run button compiles your Mission Control Sequence in the MCS. You have the choice of compiling an individualized segment or compiling all of the segments from beginning to end.

### Summary Button

The Summary button is a wealth of information that includes specific run summary data by segment. The summary reports the initial and ending states of the satellite, as well as segment-specific tasking parameters based on the attributes.

### Clear Graphics

As the Mission Control Sequence is run, by default, graphics of the outcome possibilities are also run for every iteration attempted. The Clear Graphics button clears the graphics of these previous runs from the 2D and 3D Graphic windows within the software.

## MCS Options

This button allows modifications to the visualization of the MCS, as well as calculations and the defining of user variables for further plug-in analysis. The General tab allows you to define if and how you want to visualize your graphics in the 2D and 3D Graphic window and calculate graphic updates. Additionally, setting a time tolerance for trip values and the minimum step size for ephemeris is set here. The Targeting tab allows you to define the active status for the differential correctors, set up logging, allow default, or not allow nesting target sequences. The Targeting tab also enables restoration of nominal values. The User Variables tab is used to declare variables used during propagation and finite maneuver segments. This variable must also be declared in an Equation of Motion (EOM) function plug-in. The EOM function plug-in is defined by AGI as "the derivative of four user variables: effective impulse, and the $x$-, $y$-, and $z$-components of integrated delta-V." This nomenclature for the user variable needs to be an exact character match of the names registered in the plug-in.

## Segment Properties

Segment Properties allows you to build the Mission Control Segments (MCS) using VPL techniques (see Figure 10.4). The Segment Properties option opens to the graphical command segments. Here, you select the segments and move them into the MCS to use correct components to accomplish your

**FIGURE 10.4**
Segment selection.

mission needs. Behind each graphical representation segment is a section of code that will modify the propagation of the satellite.

## MCS

The Mission Control Segments within the MCS are displayed and may be easily modified in position, color, or sets. Drag-and-drop functionality, as well as cut, paste, modify, and right-click features are easily used within the MCS. The Mission Control Segments within the MCS resemble Figure 10.5.

### Defining Mission Control Segments

Semantically, each Mission Control Segment has attributes changes for every segment. The hierarchy chart in Figure 10.6 identifies the options for the segments in the MCS.

### Initial State

This segment defines the initial conditions of your spacecraft and the coordinate system in which it is defined, and allows for custom scripting for further refinement. The default coordinate system is Earth Inertial with an array of many other coordinate systems found with the Vector Geometry Tool from Earth or other central body. Additionally, you may customize your coordinate system if desired.

The Coordinate System Type (Figure 10.7) components the Coordinate Systems (Figure 10.8) work together and are used to establish sets of libration point axes and other position points. This is one of the fundamental strengths of using Astrogator. To do this, a coordinate system is selected, than the coordinate type is configured. There have been many successful missions using the Coordinate System type and the Coordinate type to build libration points, including SOHO, ACE, and lunar orbiting missions.

**FIGURE 10.5**
Mission Control segments.

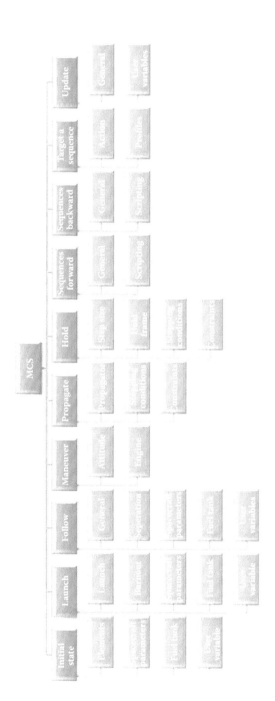

**FIGURE 10.6**
Mission Control segment hierarchy.

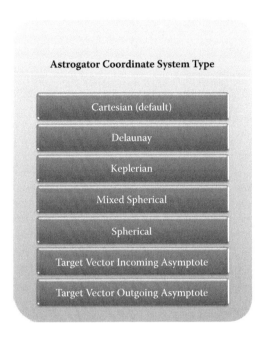

**FIGURE 10.7**
Astrogator coordinate system type.

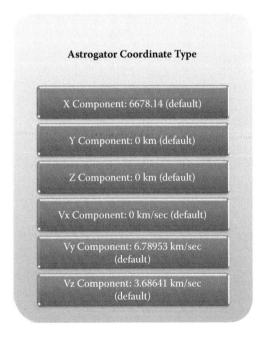

**FIGURE 10.8**
Astrogator coordinate type.

There are classes on how to use Astrogator and go in depth into the how to develop libration points into your scenario. Applied Defense Solutions Chief Scientist, John Carrico, was the lead developer of the Astrogator module and his team teaches courses on Astrogator when requested.

### Initial State Tool Segment Parameters

The Initial State Tool allows you to select the ephemeris for the New Vector Source. The default is the STK file source type, but you may also choose the following file types: Propagated Ephemeris, IIRV T9, IIRV TI, IIRV, EPV, and NASA IIRV. After the selection of the file type is complete, you will need to select a file source. The source types are Satellite (*.sa), Satellite Ephemeris (*.sae), Ephemeris (*.e), Missile (*.mi), and Launch Vehicles (*.lv). Vector Selections by default are interpolated. However, you may also choose to calculate to the closest point, closest previous, closest following, first point, or last point.

### Spacecraft Parameters and Fuel Tank

Spacecraft (see Figure 10.9) and Fuel Tank (see Figure 10.10) parameters for the initial state work together to define the spacecraft's physical values. Again, all values can be modified to make the initial state more accurate.

### Launch Segment Parameters

The Launch Segment models a simple curve based on simplified flight path methods. This is the same modeling that is used with the simplified missile object. Launch and Burnout are unique parameters to the Launch Segment; however, the Spacecraft Parameters, Fuel Tank, and User Variables information is formatted similarly to the Initial State parameters, but may hold unique values that are specific to this segment. Again, all defaults listed (see Figures 10.11, 10.12) may be modified to refine unique requirements.

Spacecraft Default Parameters

**FIGURE 10.9**
Spacecraft default parameters.

**Fuel Tank Default Parameters**

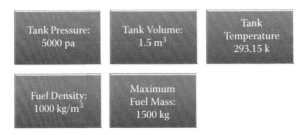

FIGURE 10.10
Fuel Tank default parameters.

**Launch Defaults**

FIGURE 10.11
Launch default.

**Burnout Defaults**

FIGURE 10.12
Burnout defaults.

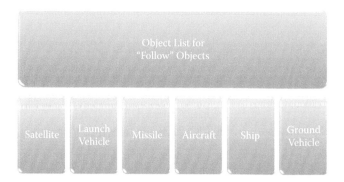

**FIGURE 10.13**
Follow segment parameters.

## Follow Segment Parameters

The Follow segment is used to launch as "attached" to another moving STK Object. In order to use this segment, you need another moving vehicle defined to use as the primary object for this spacecraft to follow. Separation parameters continue to follow the chosen moving vehicle until the spacecraft is told to stop based on your selected stopping conditions (see Figure 10.13).

## Maneuver Segment

The Maneuver segment models maneuver capabilities depending on the Attitude and Engine type you have. In the Engine tab, you define either finite or impulsive maneuver capabilities. These two maneuver types are computed completely differently. If you need to set up a maneuver type, be sure you have reviewed the Help files and have some basic understanding of the equipment you are modeling to match engine types. Without this information, your accuracy level will not be as robust.

## Hold Segment

The Hold segment pauses your vehicle by maintaining the motion pattern in a "Hold" frame for the vehicle. You define the frame in detail by step size, coordinate system, hold attitude, minimum/maximum propagation time, and the same stopping conditions that are shown in Figure 10.14.

## Propagator Segment

The Propagator segment (see Figure 10.15) gives the space vehicle motion until it meets one of the stopping conditions defined by you. These stopping conditions are the same stopping components listed previously. However,

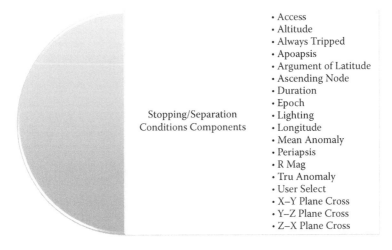

Stopping/Separation
Conditions Components

- Access
- Altitude
- Always Tripped
- Apoapsis
- Argument of Latitude
- Ascending Node
- Duration
- Epoch
- Lighting
- Longitude
- Mean Anomaly
- Periapsis
- R Mag
- Tru Anomaly
- User Select
- X–Y Plane Cross
- Y–Z Plane Cross
- Z–X Plane Cross

**FIGURE 10.14**
Stopping/Separation conditions.

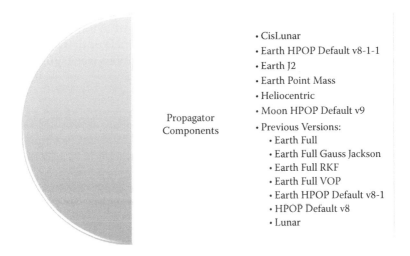

Propagator
Components

- CisLunar
- Earth HPOP Default v8-1-1
- Earth J2
- Earth Point Mass
- Heliocentric
- Moon HPOP Default v9
- Previous Versions:
  - Earth Full
  - Earth Full Gauss Jackson
  - Earth Full RKF
  - Earth Full VOP
  - Earth HPOP Default v8-1
  - HPOP Default v8
- Lunar

**FIGURE 10.15**
Propagator components.

you may select any of the other propagator components needed from the component list.

### Sequence Segment—Forward and Backward

The Sequence segment uses the power of scripting languages, either Visual Basic or JScript, to organize segments and pass the results from one segment to the next segment. Those segments identified within the

**Propagator Segment Defaults**

| | | |
|---|---|---|
| Propagator: Earth HPOP Default v81-1-1 | Stopping Conditions: Duration | Trip: 43200 sec |
| Tolerance: 1e−008 sec | Sequence: stop | Condition inherited by automatic sequences (on) |

**FIGURE 10.16**
Propagator segment defaults.

script need to be defined within the MCS. Sequencing drives the next set of segments to run until the criteria for the sequence is met. Both forward and backward sequences may be used nested with the Target Segment inside each other. These segments drive very complex maneuvers, such as station keeping.

In the example scenario found with the STK version 9.2 entitled "STK9_ Astrogator" in the Open/Online Examples section, you can explore Astrogator in a scenario to optimize a two-burn transfer. This scenario is a good example of basic forward sequencing, with the declared variables labeled in the Sequence Scripting tab (see Figure 10.17) and a sample of how to implement the scripting written in Visual Basic scripting.

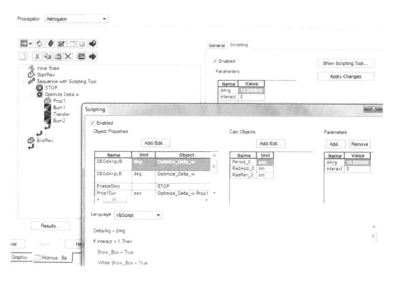

**FIGURE 10.17**
Sequence and scripting with Astrogator.

**Target Segment**

The Target segment controls the nesting functionality for the segments. The Target segment is used to customize solutions to the specific problem. For instance, if you were to create a station-keeping model, a Target sequence for each bounding criteria of your station box would be used as separate targets under a forward sequencing section, perhaps one labeled North-South Station Keeping and East-West Station Keeping. The Targeting segments would introduce nested commands to correct positions based on the scripted segment and plug-in added. In version 9.n, there is an example scenario of station keeping with the STK same scenarios that come installed with your software.

## Other Astrogator Components

Figure 10.18 through Figure 10.25 show examples of other Astrogator components.

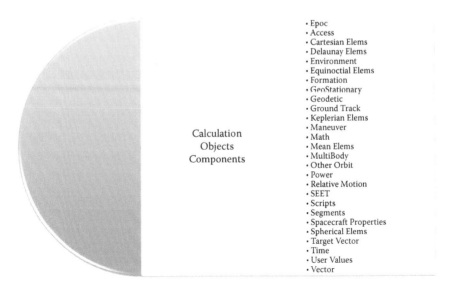

Calculation Objects Components

- Epoc
- Access
- Cartesian Elems
- Delaunay Elems
- Environment
- Equinoctial Elems
- Formation
- GeoStationary
- Geodetic
- Ground Track
- Keplerian Elems
- Maneuver
- Math
- Mean Elems
- MultiBody
- Other Orbit
- Power
- Relative Motion
- SEET
- Scripts
- Segments
- Spacecraft Properties
- Spherical Elems
- Target Vector
- Time
- User Values
- Vector

**FIGURE 10.18**
Calculation Object components.

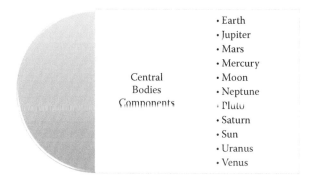

**FIGURE 10.19**
Central Bodies components.

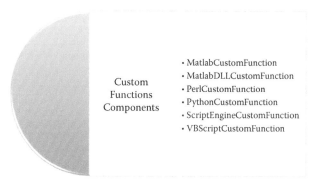

**FIGURE 10.20**
Custom Function components.

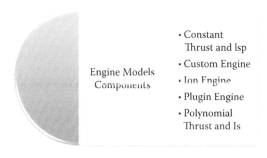

**FIGURE 10.21**
Engine Model components.

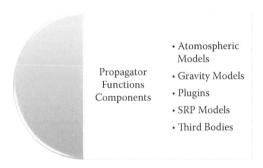

**FIGURE 10.22**
Propagator Function components.

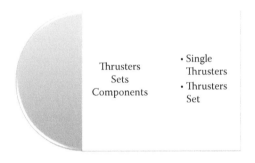

**FIGURE 10.23**
Thruster Sets components.

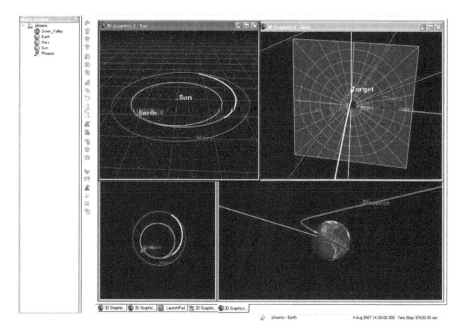

**FIGURE 10.24**
Advanced trajectory design is much easier with the Visual Programming Language used within Astrogator.

# 11

## Child Objects

---

### Objectives of This Chapter

- Define Child Object
- Attach a Child Object
- Communications Basics with the STK Child Objects
- Components Available for Communication Objects

---

### Child Objects

The child object is subordinate to another STK Object. The majority of the properties are hierarchical and inherited. The child objects require a parent or other child object as a standard STK Object to attach to. Child objects, subobjects, and attached objects in STK are synonymous and considered the same thing in STK documentation. In this book, they are referred to as the "STK Child Object." Antennas, radars, receivers, transmitters, and sensors are all forms of child objects (see Figure 11.2). These objects are primarily used in signal communication evaluations. The sensor object is used to model sensor footprints for imagery evaluations or as a gimbal for other child objects.

By default, STK Child Objects inherit many attributes from the parent object. The most important inheritance is the position of the parent object. If the parent is a fixed point object, such as a Target or Facility, the position of latitude, longitude, and altitude is the same position of the attached child. Positional inheritance is also used for moving vehicle parents. These objects inherit all velocity, positional, and orbital elements used to set up the parent object.

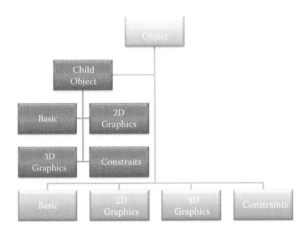

**FIGURE 11.1**
Hierarchy of the parent-child relationship.

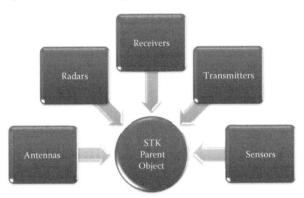

**FIGURE 11.2**
Types of Child Objects.

## STK Child Object Methods

STK Child Object methods are basically the same as those for other objects. There are four methods for each child object. They all allow for specific saved file types. The Sensor can pull in a saved *.sn file, the Antenna can pull in a *.antenna file, the Receiver object calls for a *.r file, and lastly, the Transmitter object needs a *.tx file (see Figure 11.3). The STK Child Object uses information saved within the AGI Data Federate or opens to the STK Child Objects properties page with the default child object.

There is not an actual visual 3D model for child objects; therefore, the chief object cannot actually be seen by itself within the STK environment. Their

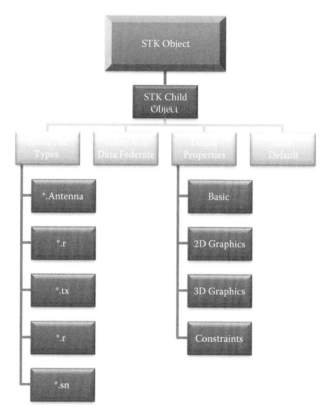

**FIGURE 11.3**
Child Object file types and attributes.

body vectors are based using a centroid or point located on the body and pro-jections from signal patterns to sensor swaths. This position may be modi-fied visually and analytically by modifying the offsets and attach points.

## Attaching a Child Object on a Parent Object

The attach point of the child object is based on the parent STK Object class. The child object default position on the fixed point is at negative zenith of the parent body. The child object's Body Zenith is at 90 degrees, where the boresight of the child object defaults. For a moving STK Object vehicle, the object will attach at +Zenith pointing toward a central body for vehicles such as aircraft, missiles, rockets, and satellites. Any attach point may modify the position for visual and calculation uses (see Figures 11.4–11.7).

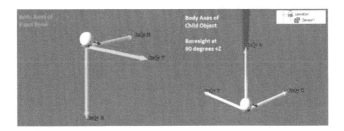

**FIGURE 11.4**
The default fixed body attachment.

**FIGURE 11.5**
The default moving vehicle attachment.

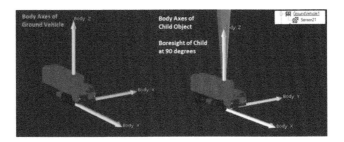

**FIGURE 11.6**
The modified moving vehicle attachment.

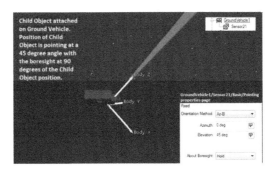

**FIGURE 11.7**
Child Object attached to a ground vehicle.

## Sensor Models

Sensors attach at the same position as the other STK Child Objects; however, they do not function the same as the Communication STK Child Objects (the antenna, transmitter, receiver, or radar). The Sensor object primarily functions in one of two ways: it models a sensor swath, or it acts as a mechanical device. The sensor swath could model a camera lens or any device needing to emulate the footprint of a satellite or other STK Object. This action from the mechanical device models emulating a gimbal for other child objects. The gimbal relationship would be another level of hierarchy—a child of a child object relationship.

The sensor swath is defined within the basic properties of the Sensor object (see Figure 11.8). The sensor has an array of sensor types found within the SensorName/Basic/Definition properties. The Simple Conic Cone Angle is 45 degrees by default for the sensor's field of view. Custom sensors can be modeled by either modifying existing sensors or bringing in a pattern file (*.pattern). The location of the sensor is determined on the Location page. Here, you can model an offset vertex, fixed displacement, or define a specific

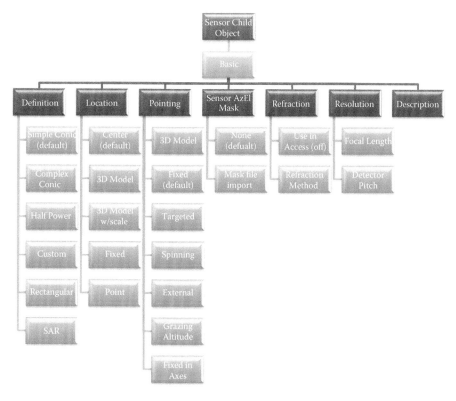

**FIGURE 11.8**
Sensor object attributes.

point where you would like to put your sensor on the parent body frame. The Pointing properties page allows you to set up the sensor as a mechanical device for slewing capabilities. The device can track, spin, graze, or use a vector-based pointing type toward a set of reference axes. Pointing capabilities are license dependent. The Resolution properties allow you to define the focal length and the detector pitch of your lens to compute the ground sample distance of the footprint. Azimuth-Elevation Masks are used frequently within the Sensor objects to model fixed-body coordinates from the parent object that emulates line-of-sight obstruction to the sensor. Synthetic Aperture Radar (SAR) simulates the field of regard of a SAR sensor from a satellite or aircraft. The Sensor object is a powerful child object used for the most simple to the more in-depth forms of modeling and simulation.

## The Child-of-a-Child Relationship

The child-of-a-child relationship is a deeper level of hierarchy and inherency. The sensor acts as the intermediate child in this relationship and models the slewing action of a gimbal. The attached child objects can attach themselves to the sensor in the exact same way as they would to a parent object (see Figure 11.9).

## Modeling Communications Equipment with STK Child Objects

Modeling basic signal communications equipment with the STK Child Object emulates both the simplistic and the robust signal evaluations. This is

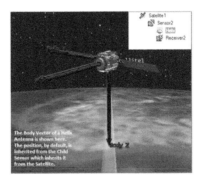

**FIGURE 11.9**
Body vector of a helix antenna on a satellite with a sensor acting as a gimbal, and a child object of a satellite.

dependent on the knowledge you have of your equipment and how deep the analysis is required. Basic signal communications with STK Child Objects model signal propagation, attenuation, degradation, link budgets, jamming, and signal strength. The STK Child Object models the equipment. The STK Tools handle the signal evaluations. Here, we cover the basics in an STK Child Object's unique attribute development using the ontological development for communications modeling and simulation within STK.

STK Child Objects used to model signal evaluations are Antennas, Transmitters, Receivers, and Radars. Signal evaluations use the method of ontological conventions by first developing STK Objects, then attaching a child object. Attribute development of the child objects further refines the quality of analysis to meet the needs of the user. Based on ontology, the relationship is developed between the setup of the object and the actual signal processing from the event detection. The output for the ontological study would be link budgets, signal evaluations, visualization of signal propagation, and pattern, as well as the line-of-sight calculations in the form of reports, graphs, and maps.

An STK Child Object models the equipment for communications; however, it does not model the signal processing handled within the event-detection tools. This is a separate part of the ontological relationships explored within STK. The properties for all STK Child Objects use modifiable components to customize signal processing within the software. Signal component models used with the STK Child Objects are Antennas, Atmospheric Absorption models, Filters, Laser Atmospheric models, Rain Loss models, Transmitters, and Receivers.

Within the Transmitters' and Receivers' properties, the STK Communications object, the Antenna components are either linked or embedded within the model properties. The type of transmitter or receiver determines if the antenna can be further defined from the standard omnidirectional default antenna used with the simple and medium models. If complex, multibeam sources are chosen, then the user can customize antenna types from a list of antenna component models.

## The Transmitter STK Child Object

The Transmitter with STK is a valuable Child Object. Like other STK objects, the object is defined within the Basic/Definition page. There are several types of models to choose from and like Astrogator, the Communications module uses STK Components models. Transmitter types give an increased capability of defining the properties and antenna:

- Simple Source
- Medium Source
- Complex Source
- Multi-Beam

Retransmitter types give an increased capability of defining the properties and antenna:

- Simple Re-Transmitter Source
- Medium Re-Transmitter Source
- Complex Re-Transmitter Source

Custom Transmitter types are for specific use, scripts, or MATLAB:

- Laser Source
- Script Plug-in RF
- Script Plug-in Laser
- GPS Satellite Transmitter

## Transmitter Properties

The Transmitter properties allow you to set up a simple transmitter when the specific details of the transmitter are not available (see Figure 11.10). Figure 11.11 shows the difference in the Model Source Types of Transmitters used with STK. The Simple and Medium Source Transmitters have options to modify basic properties. For instance, the Simple Source allows for Effective Isotropic Radiated Power (EIRP) to reflect a more evenly distributed power with a generic omnidirectional antenna without dimension and not leaving any nulls. The more specific power and antenna knowledge you have, the more flexibility for accurate modeling is available. The Medium Source Transmitter still uses the generic omnidirectional antenna; however, it allows you to set both RF power output at the antenna and the antenna's isotropic gain distribution if it is known. Complex and Multi-Beam sources allow you to select from an array of over 50 antenna components from the Components Library. Each antenna type has power, orientation, and specific setup attributes available. Like most STK Objects, the Transmitter has default parameters, as shown in Figure 11.12, for an initial set-up of the Child Object. However, these defaults are rudimentary and in order to get quality communications analysis should be refined to match your actual equipment.

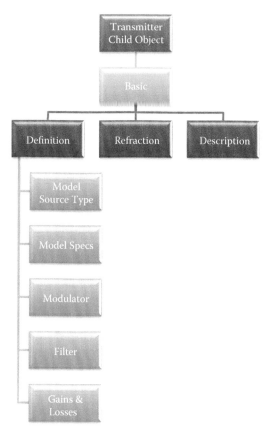

**FIGURE 11.10**
Transmitter Child Object hierarchy.

## Retransmitter Properties

STK Communication in Version 9.1 was still analyzing using received power as the primary defining point. Transmitted power did not become a robust part of the equation until later on in the version 9 development. The Re-Transmitter Child Object uses a little of both objects and is defined much the same way as the Transmitter as you need to define the type of receiver, the antenna type and refine your specifications. Figure 11.13 shows us the difference between the model types of the Re-Transmitter. In addition to the Simple, Medium and Complex model types, other types of Re-Transmitters can be Laser, Plug-in or GPS specific models that are more customized as shown in Figure 11.14. Retransmitters are similar in programming nature to a basic transmitter and receiver; however, due to the integrated capability,

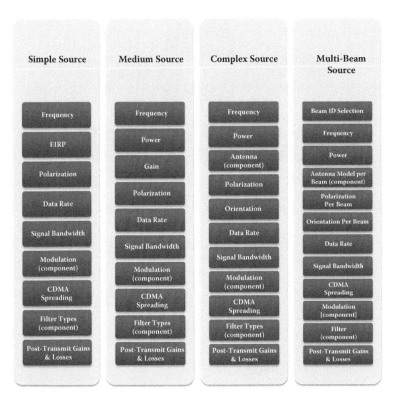

**FIGURE 11.11**
Model source types of retransmitters.

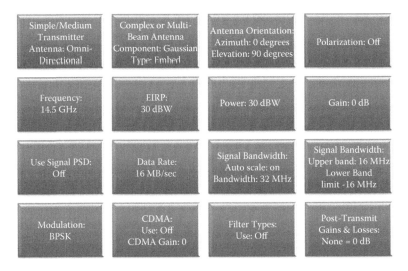

**FIGURE 11.12**
Transmitter defaults for Basic properties.

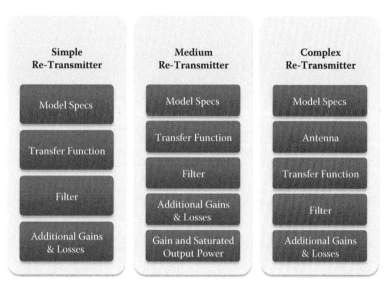

**FIGURE 11.13**
Types of retransmitters.

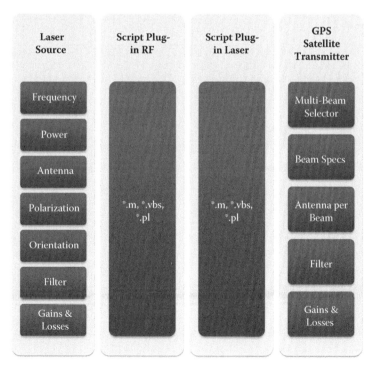

**FIGURE 11.14**
Specialized equipment.

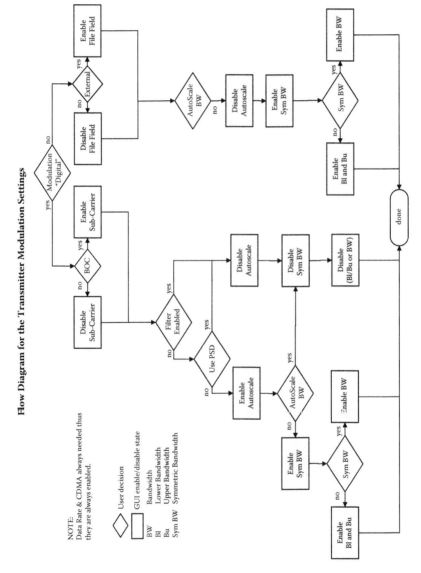

**FIGURE 11.15**
Flow diagram for the transmitter modulation.

the retransmitter has an increased complexity as it resends a modified signal after it has been received. Retransmitters work both under analog (RF) and digital communications. A combination of RF and digital transponders may be modeled in multihop evaluations.

## The Receiver STK Child Object

Receiver types give an increased capability of defining the properties and antenna

- Simple Receiver Source
- Cable Receiver
- Medium Receiver Source
- Complex Receiver Source
- Multi-Beam

## Receiver Properties

The Receiver object has properties designed to emulate receiver models. Resembling the Transmitter object, the Basic/Definition allows you to select an increasingly more complex Receiver object type. Both the Simple and Medium Receivers use a default omnidirectional antenna. The Complex

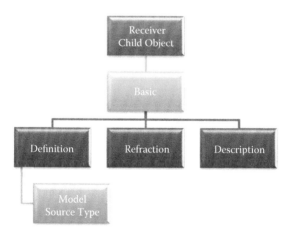

**FIGURE 11.16**
Hierarchy for receiver Child Object.

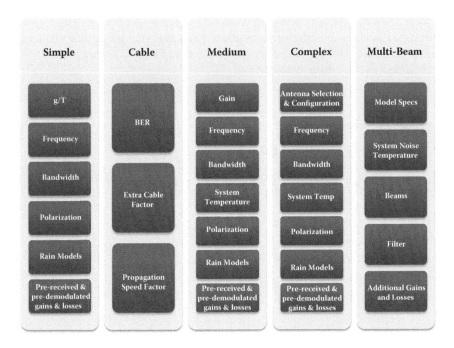

**FIGURE 11.17**
Receiver model source types.

and Multi-Beam allow you to select from the Antenna Components list. The default parameters for the Receiver Child Object model source type is set to have an omni-directional antenna, much like the default Transmitter (see Figure 11.18). Additionally, polarization values, gain, and other parameters are preset and are not designed to give more than high level analysis without refinement of the model types.

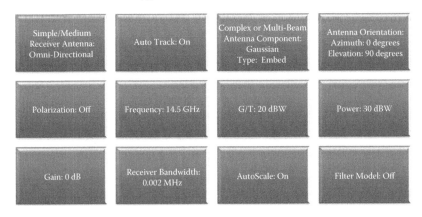

**FIGURE 11.18**
Attributes for the receiver model.

## The Antenna STK Child Object

When we select or define our Antenna type, the first option we have is define the Antenna to be either linked or embedded when used with Transmitters or Receivers. The antenna component is then defined by selecting the matching parameters to your actual equipment within the frequency, beamwidth, diameter, main-lobe gain, efficiency, and back-lobe gain. Orientation of the antenna should also be defined. The component page change and are dependent on the antenna type you select (see Figure 11.19).

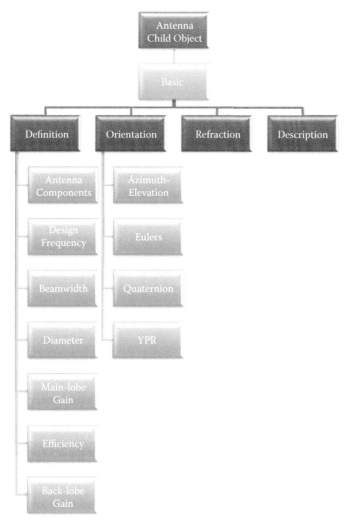

**FIGURE 11.19**
Hierarchy for Antenna Child Object.

## STK Communications Components

See Figure 11.20 through Figure 11.28 for examples of STK Communications components.

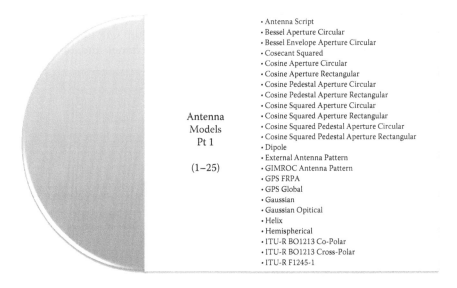

Antenna
Models
Pt 1

(1–25)

- Antenna Script
- Bessel Aperture Circular
- Bessel Envelope Aperture Circular
- Cosecant Squared
- Cosine Aperture Circular
- Cosine Aperture Rectangular
- Cosine Pedestal Aperture Circular
- Cosine Pedestal Aperture Rectangular
- Cosine Squared Aperture Circular
- Cosine Squared Aperture Rectangular
- Cosine Squared Pedestal Aperture Circular
- Cosine Squared Pedestal Aperture Rectangular
- Dipole
- External Antenna Pattern
- GIMROC Antenna Pattern
- GPS FRPA
- GPS Global
- Gaussian
- Gaussian Opitical
- Helix
- Hemispherical
- ITU-R BO1213 Co-Polar
- ITU-R BO1213 Cross-Polar
- ITU-R F1245-1

**FIGURE 11.20**
Component: Antenna Models part 1.

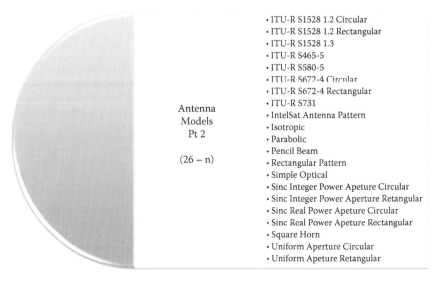

Antenna
Models
Pt 2

(26 – n)

- ITU-R S1528 1.2 Circular
- ITU-R S1528 1.2 Rectangular
- ITU-R S1528 1.3
- ITU-R S465-5
- ITU-R S580-5
- ITU-R S672-4 Circular
- ITU-R S672-4 Rectangular
- ITU-R S731
- IntelSat Antenna Pattern
- Isotropic
- Parabolic
- Pencil Beam
- Rectangular Pattern
- Simple Optical
- Sinc Integer Power Apeture Circular
- Sinc Integer Power Aperture Retangular
- Sinc Real Power Apeture Circular
- Sinc Real Power Apeture Rectangular
- Square Horn
- Uniform Aperture Circular
- Uniform Apeture Retangular

**FIGURE 11.21**
Component: Antenna Models part 2.

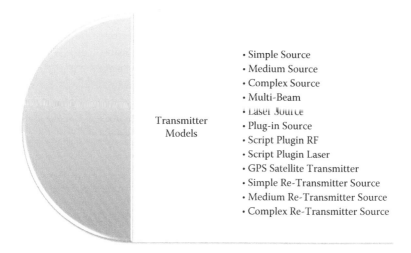

**FIGURE 11.22**
Component: Transmitter Models.

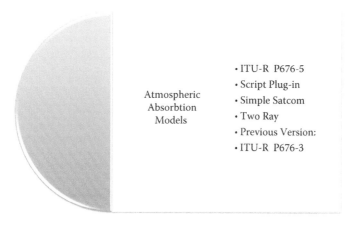

**FIGURE 11.23**
Component: Atmospheric Absorption Models.

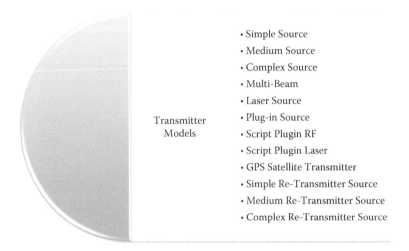

- Simple Source
- Medium Source
- Complex Source
- Multi-Beam
- Laser Source
- Plug-in Source
- Script Plugin RF
- Script Plugin Laser
- GPS Satellite Transmitter
- Simple Re-Transmitter Source
- Medium Re-Transmitter Source
- Complex Re-Transmitter Source

Transmitter Models

**FIGURE 11.24**
Component: Transmitter Models.

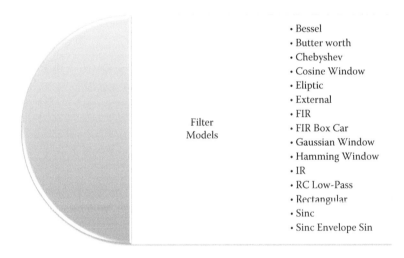

- Bessel
- Butter worth
- Chebyshev
- Cosine Window
- Eliptic
- External
- FIR
- FIR Box Car
- Gaussian Window
- Hamming Window
- IR
- RC Low-Pass
- Rectangular
- Sinc
- Sinc Envelope Sin

Filter Models

**FIGURE 11.25**
Component: Filter Models.

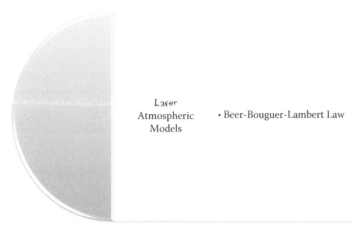

**FIGURE 11.26**
Component: Laser Atmospheric Model.

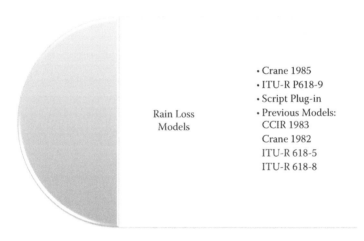

**FIGURE 11.27**
Component: Rain Loss Models.

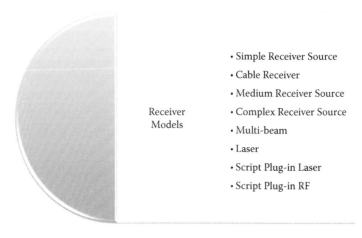

Receiver
Models

- Simple Receiver Source
- Cable Receiver
- Medium Receiver Source
- Complex Receiver Source
- Multi-beam
- Laser
- Script Plug-in Laser
- Script Plug-in RF

**FIGURE 11.28**
Component: Receiver Models.

# 12

## Constellations

### Objectives of This Chapter

- Define a Constellation Object
- Constellation Constraints

### The Constellation Object

A Constellation is an array of similar STK Objects. The Constellation stores information for multiple STK Objects of the same object type. The array is an array-like object used for analysis in place of using individual objectsin an empty header. The purpose of using an STK Constellation over an individual list of STK Objects is the ability to analyze pair-wise comparisons between objects. Here, object type attributes as well as specific Constellation constraints are considered while evaluating the relationships during event detection using event-detection tools. The Constellation allows each object within the array to fully utilize properties and constraints for these stored objects. In addition, STK Constellation has unique object attributes and constraints to consider. For instance, you would use it to analyze a constellation of Iridium satellites in a constellation to see if you could have signal with a ground network of satellite phone receivers used for coordinating special events.

### Filling a Constellation

Constellations are created by assigning already existing STK Objects that are listed within the Object browser to the "Assigned Objects" section within the STK Constellation Object interface. The Constellation reads all available STK Objects and then allows you to select the ones you need to fill the

The Ontological Relationship:
defined objects and tools

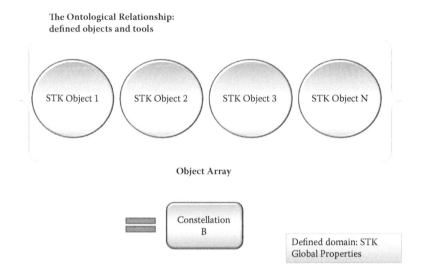

FIGURE 12.1
A Constellation is a group of STK Objects. In order to use the constellation as an object for analysis, it needs to be filled with "assigned" objects much like an array in computer programming.

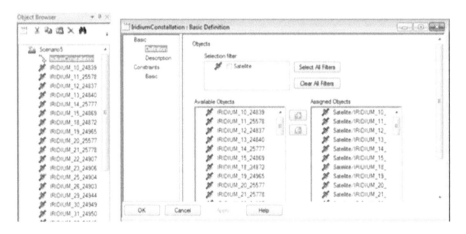

FIGURE 12.2
The STK GUI defining an STK Object.

Constellation array. As you fill the STK Constellation array with the assigned STK Objects, you can further define the constellation by selecting specialized constraints. When analyzing a chain access using a constellation without constraints set, as the Figure 12.4 shows, the STK Object will select each item within the constellation and consider it to see if it has a successful access or not. The constellation allows you to have increasingly complex analysis in an organized and logical manner. When we add in constraints, the analysis will become even more complex.

**FIGURE 12.3**
Constraints of a Constellation. These constraints help you limit the analysis parameters in the use of the constellation.

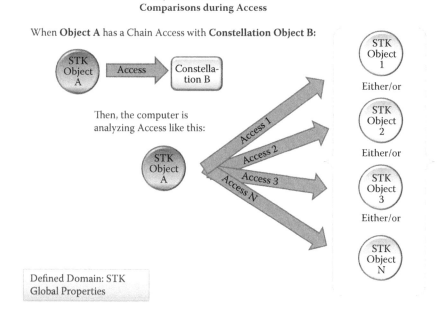

**FIGURE 12.4**
A single Constellation Chain Access event is similar to a 1:M Access event.

## Constellation Constraint Conditions

Constraints allow you to refine the definition of a Constellation array. Pair-wise comparisons are established between an object and Constellations—and sometimes even between Constellations—by the ontological use of the appropriate STK Tool.

## All Of

This constraint requires that all STK Objects assigned to the Constellation as an array have a valid Access when used. In Figure 12.5, you have a Chain Access event between STK Object A and Constellation Object B. If we were considering the use of four objects within Constellation B, then the constraint "All of" would require that all four STK Objects would need to have Access to STK Object A to be valid.

## Any Of

"Any of" is the default constraint for a Constellation array. This allows a valid Access if one or more of the STK Objects meets the Access requirements defined by the STK Objects, the STK Tool, and the Constellation Object. As shown in our first constellation example in Figure 12.4, we would again consider a Chain Access between Constellation B and STK Object A. An event detection would be valid if any STK Object that was assigned within Constellation B was to have Access with STK Object A.

## At Least N

The STK Objects inside the array must meet the minimum requirements at number value N Objects for a valid Access. For instance, we have selected 12 Satellites in our Constellation array, but to meet the requirements for a valid

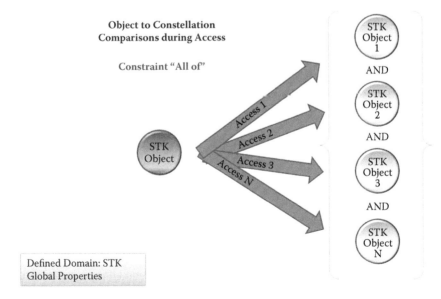

FIGURE 12.5
The "All Of" constraint defined.

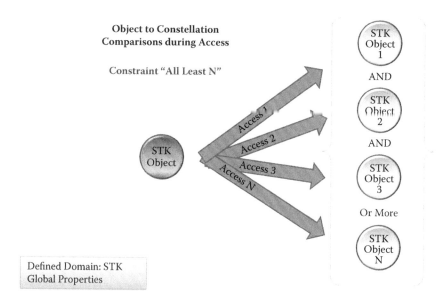

**FIGURE 12.6**
The "At Least N" constraint defined.

position on our STK Object A we will need a minimum of two satellites with valid Access to Object A at all times for redundancy. More Satellites would allow for more accurate data feeds, but two is the minimum required for our exercise. We would select the constraint of At Least N and give the *N* value the number 2. This would mean that in order to have a valid Access from Constellation B, Satellites 1 and 2 (or more satellites) within the array would need to have a valid Access with STK Object A.

## Exactly N

In Figure 12.7, we consider the "Exactly N" means that you can select the exact number of assets within your Constellation array to maintain Access to your STK Object A at all times to have a valid Chain Access. For instance, you have a requirement to have two Ground Stations tracking a launching rocket at all times while in flight. More Ground Stations would be cost prohibitive and less would leave the rocket vulnerable.

## None Of

The "None of" constraint means when there is no intervisibility, then the results you are looking for are valid. Here, you would be looking for results

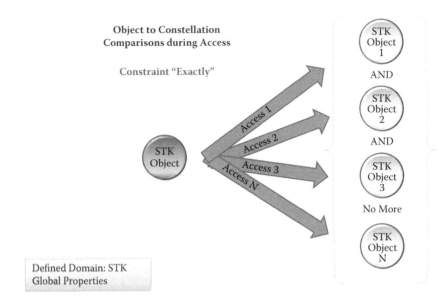

**FIGURE 12.7**
The "Exactly" constraint defined. In this configuration, only the number of accesses defined are allowed. If there are more or less access intervisibility connections, then the constraint would be invalid and the access would read as false.

when you don't have any intervisibility between the constellation and the object rather than when you do have Access. What is different with using the "None of" constraint is that a Constellation using this constraint cannot be the first object in the Chain—it has to be at least in the secondary node or beyond.

## Cross Parent

The Cross Parent Access constraint is on by default and is primarily used with child objects. Cross Parent Access allows child objects (sensors, transmitters, receivers, etc.) to link to other child objects on differing parents within the Constellation array. However, when considering signal communications, this can cause link budget miscalculations if left enabled. Best practices indicate that in up-link and down-link modeling and when computation is more accurate, it is best to make sure you have disabled the Cross Parent constraint. This requires that both the up-link and the down-link communications reference the same parent object.

# Section III

# STK Tools

# 13

---

## *STK Tools*

---

---

**Objectives of This Chapter**

- Tools and Ontology
- Tool Classes
- Time Computations
- Understand Light Time Delay

---

## Tools within STK

Tools are analytical object classes inside STK that leverage computational and analytical functionality by forming relationships between STK Objects. Tools are not like the physical objects as found in the Objects section of the book but make up an object class that develops calculations between two or more objects. These object relationship algorithms were developed using forms of event detection, signal analysis, or positional evaluations that are eventually handled within the STK Engine. Because of the differences in functionality and to lessen confusion, we address the tool object class simply as an STK Tool.

The ontological relationships are defined by STK Tools. They have basic default attributes. These attributes may be customized within the tools or by adding additional data providers to refine the details prior to calculating the output. Here, we use the formal ontological study of STK to understand how the attributes of object-oriented STK Objects, Tools, and Outputs are all dynamically interactive. As the relationship, or the ontology, among the tools and objects is defined, the STK computational engine analyzes these relationships. The engine reads the algorithms and creates the output in the format the user defines as maps, pictures, or graphs. The data providers allow the user to select what is represented and how.

Currently, inside the STK interface, tools appear in multiple places. They are listed in the STK Tools task icon menu, the Utilities task menu, inside

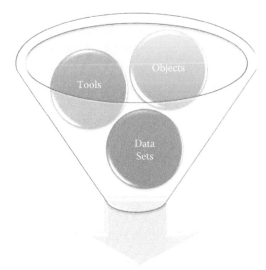

Maps, Reports, Graphs
& Data Providers

**FIGURE 13.1**
Tools are needed to build relationships and give an output in STK.

the "Insert NEW STK Object" wizard, and by right-clicking the STK Objects within the Object browser. It is the functionality of tools that defines the relationships between objects. Tools calculate line of sight, proximity, and quality of communications signal. Tools can also define conjunction analysis and the vector geometry used on an object. Tools evaluate event-detection relationships between objects.

## Tools and Ontologies

The type of tool selected is the initial defining point of the ontology between STK Objects. Since ontology is the theoretical study of objects and their relationships, a further study of what relationship is formed when using an STK Tool is essential.

The first look of ontology is when we begin to take a deep look at the object itself. The object is defined by attributes and constraints. When we use objects within STK, the analytical strength of the software is not apparent until we apply the tools and define the relationships between and among the objects. It is during the process of defining the STK Objects and defining the STK Tools that the analysis algorithms are established. After the formulas are defined, then STK uses the STK Engine to compute the defined relationships

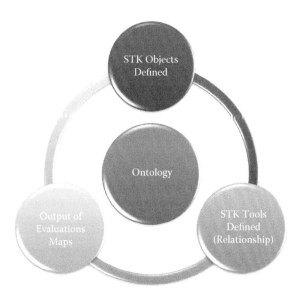

**FIGURE 13.2**
Tools are a necessary part of the STK ontology.

between the objects. These STK Tools are the foundation for the ontology as it defines the computational base algorithm of the objects' relationships.

The ontological relationships that are formed with STK Tools can be both geometric and topological. Geometric relationships have a focus on the distance between objects based on the coordinate systems and ellipsoid. Topological relationships are less concerned with the coordinate system used but are more focused on the relationships of connectivity using a dependent but separate level of evaluations, such as communications. The two relationships, both geometric and topological, are handled by different data providers within the tool.

Geometric relationships in STK consider the position and movement of the STK Objects on the ellipsoid. As objects move in position, their distance or proximity may change from one another. This may cause the objects to either establish or discontinue their ability to maintain line-of-sight computations for various reasons. STK continues to monitor the relationship based on the time intervals selected to see if Access, or some form of intervisibility, is maintained or regained. It is important to remember that many parameters of the STK Object, STK Tool, and the time intervals are user defined. Movement of any object causes fundamental spatial changes that could dynamically change the validity of intervisibility based on the definitions of the STK Object attributes and constraints, the STK Tools definitions, and the root-level Scenario definitions.

Topological relationships are formed after the geometric relationships have been established. After the STK Tool intervisibility has been established, then if there are STK Communication Objects involved, a secondary level of connections is then evaluated. Communications Objects are typically child objects. The STK Objects that specialize in communications are antennas, receivers, transmitters, and radars. Signal Communications within STK also use an STK Tool called CommSystems. This tool is used in addition to the Access, Chains, or Coverage for signal evaluations for topological ontology. Although topological relationships require a valid geometric relationship, geometric relationships do not depend on a topological relationship. Therefore, if an object does not establish Access, a signal evaluation cannot be computed. However, if STK Objects can evaluate a signal, then the objects must also have intervisibility. In addition, a signal link can never be established with STK Objects that do not use STK Communications Objects.

With STK, the geometric relationship would need to be formed before the topological relationship could be considered. However, the differences between establishing the geometric and topological relationships are not apparent to the user from the engineering workflow of the interface. STK handles this during the computational evaluation within the STK Tool. For instance, if a user wants to establish a signal link between an antenna transmitter and a receiver, he or she would only need to use the Access tool and select the "Link Budget" button to be able to begin to evaluate the link between the two communications objects. Within the Access tool algorithm, when calculating for a link budget, the tool will first set up the analysis for intervisibility and then begin to formulate algorithms for link budget signal analysis.

When we look at STK Tools, we consider the base form of intervisibility between objects. This form of ontology considers the spatial objects, their positions, and their identities. The STK Tools analyze the spatial relationships defined by topological, directional, and distance calculations that can be quantitative or qualitative. These tools can be refined in their definitions to further the ontological analysis between the objects as needed. With each modification of attribute and constraint with an STK Object or a modification of the STK Tool, a new output is presented to the user, allowing robust ontological studies to be formed within the STK environment.

## Tool Classes

STK has an array of tools. Most of these tools define the ontology of the objects. However, there are some tools that are not ontological. The ontological tools define a different relationship between the objects within STK. These tools are the Access, Deck Access, Chains, Coverage, Figure of Merit, Conjunction Analysis, Time, and Calculation tools. AGI has announced they are working on a future Volumetric tool. Nonontological tools are the Vector Geometry, Terrain, and Globe Manager tools. These tools do not define the relationship between the objects, but rather modify and leverage the objects to refine other ontological studies.

## Ontological Tools

The ontological tools focus on different ontological studies and provide a unique type of output based on the relationship defined. STK Objects may have many different forms of ontological relationships defined for the same time interval.

Access and Deck Access tools analyze intervisibility from one STK Object to another.

- Chains allow for multiple Access computations using many STK Objects in a string of valid visible objects.
- Coverage is a qualitative and quantitive analysis used abundantly for a special object called an Area Target, which is an area of interest. Coverage can also analyze a path or an attitude around an object. The Figure of Merit further refines the qualitative computation and allows the user to visualize the relationship analysis.
- The Time Tool and the Calculation Tool, which are new in version 10, allow the user to define the tool elements for Access computation at the attribute level using data providers.

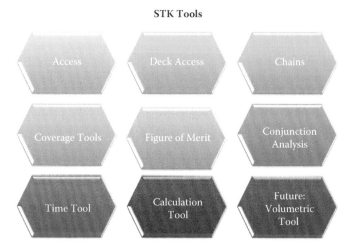

**STK Tools**

**FIGURE 13.3**
Different types of ontological tools.

- Conjunction Analysis computes the distance relationships among spatial objects as they are in motion. Computation of the distance between objects and a threshold around the objects are defined for possible intersection or conjunction.

## Nonontological Tools

As stated, there are some tools that are nonontological.

- The Vector Geometry Tool (VGT) defines positional vectors on the STK Objects. This can be used to compare these objects to the Vector Geometry of the coordinate frames of other objects or central bodies. Vector Geometry may be a defined STK Object attribute or it can be used as an STK Object constraint tool. It may also be specifically referenced with STK Tools for spatial measurements.
- Neither the Terrain Conversion Tool nor the Globe Manager Tool are used overtly ontological tools but are tools to refine the scenario-level attributes for topology and the 3D Graphics window. Although these tools affect the overt output of values, the primary use of these tools is that they become part of the definition of the root of the scenario. This is referenced by the STK Object to refine spatial position.

## Time Computations

Time computations within STK show intervals of satisfaction or validated line-of-sight analysis with the use of constraints. Most STK Tools use time as a dynamic integration within their algorithms; however, the primary tools that leverage time intervals and constraints are Access, Deck Access, and Chains.

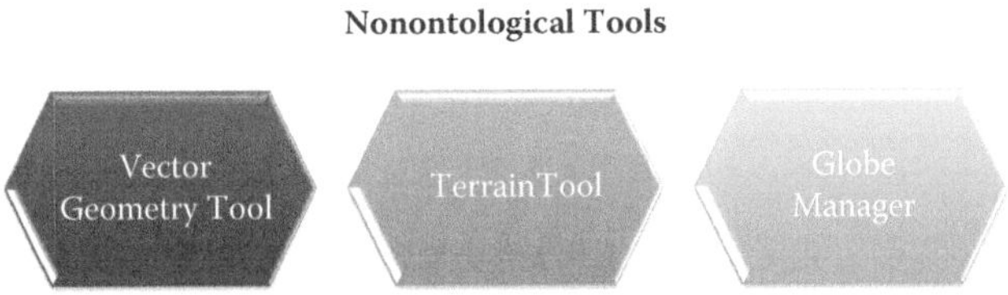

**Nonontological Tools**

**FIGURE 13.4**
Nonontological tools.

What Access, Deck Access, and Chains are looking for is a valid line-of-sight Access relationship between two or more objects along with the considerations of the defined constraints. Here, Access is the underlying algorithm used in Deck Access and Chains. For instance, a satellite receiver will receive telemetry and signal from a satellite only if the signal is uninterrupted by the curve of the Earth, or impedances from physical objects such as mountains or buildings. Although other issues, like signal jamming and attenuation, might diminish the communications, the first set of impedances will overtly cause an invalid line-of-sight connection for Access. If the relationship between the objects cannot be established, then the objects do not have valid Access, Deck Access, or Chain computation capabilities. During the time intervals that the relationship is true, this is considered valid Access. As STK is time-dynamic software, time is relevant in most calculations.

## Understanding Light Time Delay

Light Time Delay (LTD) is the measurement of time between two objects. For many objects that are relatively close in proximity, as well as most calculations on a single planetary area, the minute calculations of time delay are negligible. When dealing with objects in space and larger distances, LTD is usually necessary to consider. For this reason, Light Time Delay calculates by default within STK. There are times, however, when LTD is not needed or not specifically wanted in these larger distance calculations. These instances would be defined by the parameters within a user's analysis. When LTD is toggled off, the event detection of access only calculates using the speed of light from two objects as an infinite computation regardless of position. You can find more specific information on LTD in Appendix B.

# 14

## Access and Deck Access

### Objectives of This Chapter

- Access
- Access Defaults and Options
- Deck Access
- Deck Access Defaults and Options

### General Description

Access and Deck Access are event-detection tools that measure the intervisibility between STK Objects. Although objects have spatial definition and orientation, they do not form relationships to evaluate distances, line of sight, or signal communications. Tools evaluate these relationships. Access and Deck Access evaluate these relationships by validating if a detected connection can be made between the objects. From a spatial geometric perspective, either Access or Deck Access may be calculated from any STK Object, including a facility, target, vehicle, area target, signal device, or central body. The Access algorithm is used as the underlying structure for many STK Tools; therefore, it is the most important and basic ontological relationship to understand.

Computational considerations for Access are based on the properties of the STK Objects, the constraints defined within the objects, the orientation of the framework of the objects, time intervals set, and effects of refraction, as well as light time delay. Once either Access or Deck Access has been computed, it will continue analyzing the event-detection relationship during the time interval selected for either of the defining objects until one or more of the objects has been modified. At this point, the relationship of Access will either be recomputed or removed. When Access is computed based on the STK Objects that were selected, an output in the form of maps, reports, or graphs of the Access details can be generated.

## Access

Access is an intervisibility analysis tool used within STK. During an Access evaluation, STK Objects, either static or in motion, are compared with each other using pair-wise comparisons to see if they have line-of-sight capabilities. Many times, the STK Object is referencing a central body such as Earth's ellipsoid. Access can evaluate not only geometric ontologies of direction, distance, and spatial orientation, but also topological ontologies for signal evaluations. Access has two data relationship possibilities: a one-to-one relationship (see Figure 14.1) and a one-to-many relationship (see Figure 14.2). These relations are the foundation for the pair-wise comparisons used when evaluating Access.

### Access Logic

Access logic uses pair-wise comparisons and Boolean logic to formulate the foundation of Access calculations. This allows systematic validation for each comparison string to see if the initial STK Object (Object A) would have intervisibility to the next STK Object selected. The initial object, or object of the originating evaluation, is always singular. The STK Object that is secondary may be singular or there may be a short array of more than one.

The Access tool is computing when the destination object may be "seen" or a communication link can be established. In order for Access to be calculated, there must be intervisibility between the entities. When we look at the logical statement for Access (see Figure 14.3), it considers each STK Object and a Boolean; for instance, Access is valid if

entity A AND entity B = True

**Access**

When **Object A** has Access with **Object B:**

This is a 1:1 Relationship

Defined Domain: STK
Global Properties

**FIGURE 14.1**
A simple Access relationship.

**Many Objects and Access**

When **Object A** has Access with **many Objects**:

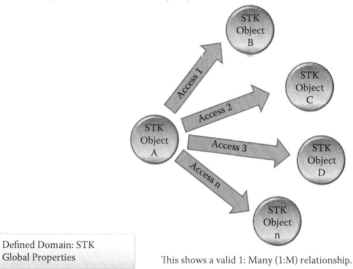

Defined Domain: STK
Global Properties

This shows a valid 1: Many (1:M) relationship.

**FIGURE 14.2**
A 1:M Access relationship.

**Access Logic**

When **Object A** has Access with **many Objects**
it need to establish at least one valid path:

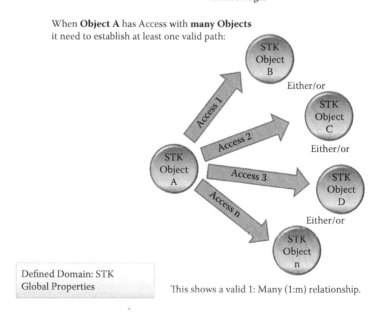

Defined Domain: STK
Global Properties

This shows a valid 1: Many (1:m) relationship.

**FIGURE 14.3**
Access logic.

If there is more than one entity for the destination objects, then the logic is valid if there is intervisibility between

entity A AND B either or A AND C either or A AND D either or

until the number of destination objects that were selected is evaluated where

(A AND B) either or (A AND C) … (A AND $n$) = True

## Access Using Geometric Ontology

The first primary form of ontological analysis is geometric in format. It is calculated by evaluating the distance between two points determined when line of sight is valid and has not been lost. When an object can visualize another object without obstruction, it is said to have intervisibility or "Access" in the terms of STK. With standard line-of-sight analysis, barring any atmospheric refraction, distance calculations are derived from the horizontal positions from the Earth—from the observer to the object being viewed—using simple trigonometric identities and functions. In a simplified manner, Access is calculated when it is possible to view objects in a straight line from one point to another (see Figure 14.4).

To fully understand this, the horizontal position must first be determined. There are primarily two different tangents off the curvature of a central body (either real or fictitious) that would define the position of the first object within an Access calculation. An apparent or True Horizon $d_t$ would be the horizontal line tangent to the arc of the central body. The Astronomical Horizon $d_a$ is the line based off the altitude of the initial object's actual position that runs parallel to the True Horizon. If either, or both, objects within the Access statement are in motion over a given time interval, Access will continue to compute through the time interval and evaluate the change in position when intervisibility is valid or invalid.

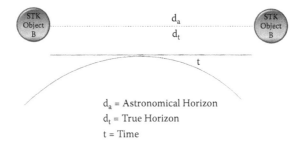

$d_a$ = Astronomical Horizon
$d_t$ = True Horizon
$t$ = Time

**FIGURE 14.4**
Access line of sight.

The calculation for intervisibility (still without considering refraction) based off the Astronomical Horizon of the object's position, considers the curvature of the central body; for example, in the case of the Earth, we would consider the surface of an ellipsoid. Initial calculations evaluate the altitude of a vehicle object based on the object's position with reference to Earth's globe surface at mean sea level (MSL), at the central body's reference ellipsoid WGS84 (World Geodetic System, 1984), or with topographical features included in Terrain files.

## Access Defaults and Options

Access defaults and options are set both globally and locally with an STK Scenario. Access computes using scenario- or root-level parameters defined from the object properties and within the Access tool. These defaults are global settings for all Access calculations used within a scenario. Global settings are considered tool attributes, whereas local options are constraints of the tool. Local Access options are found on the Access Tool panel. These can be modified for each unique Access calculation defined within a scenario.

### Access Global Defaults

Default parameters consider the time step size, time convergence based on the Access intervals, value tolerances, light time delay, light time delay convergence, and aberration type. These default parameters may be found and edited from the Edit—>Preferences/Access Defaults page in version 9.*n*. In previous versions (see Figure 14.5), this was in the Options dialog box. Star objects are an exception to the modification of the defaults. All of the following definitions can be found in the Help files section of the STK software:

*Maximum and Minimum Step Sizes* are limits allowed to elapse between samplings of the constraint function during computation.

*Time Convergence* options set the time tolerance of the start and stop times within the access algorithm.

*Value Tolerance* controls the event detection based on the constraint value's relative difference compared to the previous sample and also as an absolute, where the values are near zero based on the value minus the previous value within the tolerance.

*Light Time Delay* (LTD) is always toggled on as a default. However, there are times that LTD is not necessary or would interfere with correct analysis. When LTD is toggled in the off position, then the algorithm for access will use the speed of light to calculate Access. Access also considers the STK Object constraint of LTD. This is also usually on by default. The order of operations that STK performs calculations would be first to define Light Time Delay, then refraction, and lastly aberration. Total stellar aberration is still used in all calculations even if the *Aberration Type* and *Use Light Time Delay* modifications have been established. To learn more about how Light Time Delay is calculated, see Appendix B for the AGI document "Light Time Delay and Apparent Position."

All objects considered within the Access event are influenced by either the default parameters or the modified properties of the STK Object. Default parameters are designed to analyze the most simplified analysis in more generalized terms. Customizing the defaults to mirror real-life values increases the integrity of the analysis output. Each object has unique parameter settings and customizations. If the user has established a defined customized default, these new parameters may be saved as new "default values" from the Options

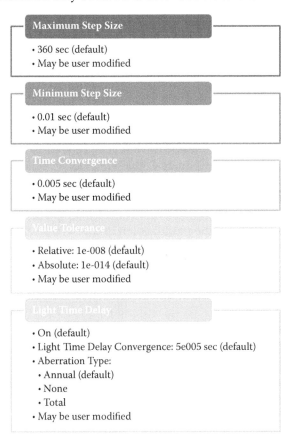

**Maximum Step Size**
- 360 sec (default)
- May be user modified

**Minimum Step Size**
- 0.01 sec (default)
- May be user modified

**Time Convergence**
- 0.005 sec (default)
- May be user modified

**Value Tolerance**
- Relative: 1e-008 (default)
- Absolute: 1e-014 (default)
- May be user modified

**Light Time Delay**
- On (default)
- Light Time Delay Convergence: 5e005 sec (default)
- Aberration Type:
  - Annual (default)
  - None
  - Total
- May be user modified

**FIGURE 14.5**
Access global defaults.

menu of the software. If a new default is desired, the user must make the necessary changes to the commercial-off-the-shelf (COTS) default parameters and then select the "Set as Default" toggle option in the bottom of the Preferences page. These changes will be saved for the current scenario only.

## Local Access Options

Access options are modifiable settings based on the Access tool dialog box found within the STK Access tool. These defaults and options are local constraint parameters used for each unique Access computation (see Figure 14.6).

*Graphics:* The Graphics constraint options define the visual display on how a valid Access appears in both the 2D and 3D Graphic windows. The default allows all the settings to be inherited directly from the Access settings found in the path selected by right-clicking the Scenario name and opening the Properties to: [ScenarioName]/Properties/2D Graphics/Global Attributes. Here, global settings on how STK Objects and mapping displays are set. By inheriting these properties locally, it leaves all Access visualization standardized. However, you may customize the visualization locally by selecting the Show Line, Animate Highlight, or Static Highlight properties.

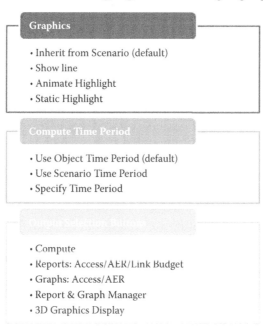

**Graphics**

- Inherit from Scenario (default)
- Show line
- Animate Highlight
- Static Highlight

**Compute Time Period**

- Use Object Time Period (default)
- Use Scenario Time Period
- Specify Time Period

**Output Selection Buttons**

- Compute
- Reports: Access/AER/Link Budget
- Graphs: Access/AER
- Report & Graph Manager
- 3D Graphics Display

**FIGURE 14.6**
Access Local defaults/options.

*Compute Time Period:* Access is calculated over a period of time and checked for valid Access times based on the interval level selected. The default settings are based on the time settings found on the STK Objects used for the Access calculations. The smallest time interval is selected for these calculations. However, there are times when you would like to evaluate Access over the full duration of the scenario or a given set of time. These are user options that may be defined on the Access panel.

*Output Selection Buttons:* The Output buttons allow you to select how you will use the information for analysis. More details about Output are discussed in Chapter 20 of this book.

- *Compute:* The Compute button simply allows the STK Engine to execute the Access algorithm. Output displays show visually on the 2D Graphic map based on the graphics display chosen at the top of the page. 3D Graphic displays display the same way if they are inherited from the 2D Graphics attributes. Typically, Access is denoted on the maps as a hard bold line.

- *Reports:* The Reports button computes the Access ontology of the objects and delivers an output in the form of a report. The Report button also generates the same visual displays that are delivered when the Compute button is selected. There is no reason to hit the Compute button prior to running a report, as the report button activates the same "Compute" commands.

    The default Access report reveals the Access times and the duration of the Access. The Azimuth-Elevation-Range (AER) report defines the Azimuth Angle, the Elevation Angle, the Range distance, and the time interval that the Access is valid based on the STK Object that is initializing the Access. This object is also known as the "From" STK Object. All computations are based on the local coordinate system of the initial object. The Link Budget button is only available when Access is defined using two or more STK Communications Objects. Antennas, transmitters, receivers, and radars have attributes and functions to define a signal and evaluate it based on standard signal communications understanding.

*Graphs:* The Graphs button generates both a standard report of the time and duration Access is valid and an AER report. Much like the Reports button, this computes the Access equation developed within the ontology using the selected STK Objects in the STK Engine. It then creates a visual graph based on the global defaults. Locally, modifications may be made on the Access graph.

*Report and Graph Manager:* The Report and Graph Manager button allows for more report options that are generated from the data providers found in the STK Access tool and STK Objects. There is a selection of premade graphs and reports, dynamic displays, and strip chart templates. In addition, there is also the ability to completely customize each graph and report by selecting the desired data providers.

*3D Graphics Display:* The 3D Graphics Display button allows you to display specific information dynamically on the 3D graphic window.

## Advanced Local Access Options

Refinements to the computation of events are handled with the Advanced Options button. This button allows modification of event detection, step size control, light time delay, and signal path options. These changes override the Access defaults from the Preferences section with the exception of stellar aberration used with Star Objects (see Figure 14.7).

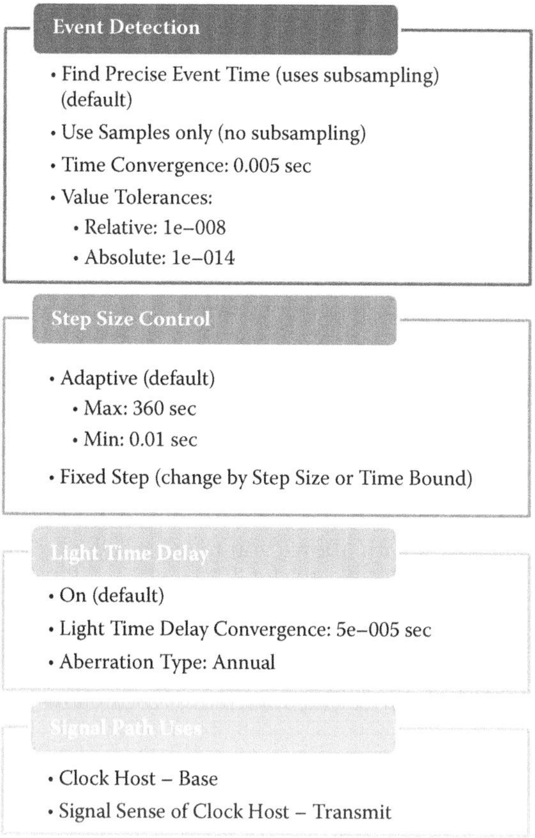

**Event Detection**

- Find Precise Event Time (uses subsampling) (default)
- Use Samples only (no subsampling)
- Time Convergence: 0.005 sec
- Value Tolerances:
  - Relative: 1e−008
  - Absolute: 1e−014

**Step Size Control**

- Adaptive (default)
  - Max: 360 sec
  - Min: 0.01 sec
- Fixed Step (change by Step Size or Time Bound)

**Light Time Delay**

- On (default)
- Light Time Delay Convergence: 5e−005 sec
- Aberration Type: Annual

**Signal Path Uses**

- Clock Host – Base
- Signal Sense of Clock Host – Transmit

**FIGURE 14.7**
Advanced Local options with default parameters.

*Event Detection:* The event-detection sampling rate is set to find precise times by subsampling the intervisibility intervals, also known as the subsampling of the step size. The sampling rate is established within the start time and stop time intervals within the STK Object. STK uses a time convergence of 0.005 sec for the time intervals between the subsampling rates. For most calculations, this should give optimum results versus the cost of time to compute. There are times, such as when dealing with space operations, when we would need to sample time over days or weeks, or when comparing the Animation Time or the Scenario Time over six months or a year. In addition, there are times when you might desire tighter sampling rates of time much smaller than a minute and choose to disregard the cost of the time it takes to calculate the response. When you consider the details of sampling for your event detection, the cost factor of calculation speed should be considered. The primary thing to remember is that the higher fidelity of time you select for convergence and the less tolerance selected for event differences, the more time it will take your computer to calculate.

*Step Size Control:* The Step Size Control is either Adaptive or Fixed. The Adaptive Step computes the upcoming step sizes using complex algorithms based on the movement of the objects and the constraints each object has defined. Fixed Step Size computes a consistent sampling size. Here, the computation is smooth and even, due to its disregard of STK Object motion or behavior in the calculations.

*Light Time Delay:* Light Time Delay is on by default. It considers aberration when it is toggled on and does not consider aberration when it is off. Calculation within the STK Engine computes, in order, LTD, then refraction, and lastly it consider any aberration computations. Refraction considers the bending of the signal path, even beyond the line of sight. Aberration uses three primary types: Total, Annual (default), or none. The Annual Aberration considers the motion of the planet or central body of reference but disregards the velocity. Total Aberration considers all motion and velocity for full computation.

*Signal Path:* The Signal Path allows you to select which STK Object has the Clock Host associated with it for signal orientation and sets the direction of signal.

## Access Data Providers

Data providers are the segments of data from the semantic level of the STK Object or STK Tool. The results of Access calculations give us information from the initial object being analyzed. This information is based on the default data providers built into the Access tool and the STK Objects used within the calculations. These default data providers, after computing with the STK Engine, give valuable quantitative information regarding the relationship between STK Objects: Azimuth, Elevation, Range, Date, and Length of Time of the Access event.

The Azimuth (see Figure 14.8) may be defined as the clockwise measurement of angular distance within a celestial sphere. It is measured as a bearing, from a north point of 0 degrees to a south point of 180 degrees along the astronomical horizon to the great circle as it intersects with the astronomical zenith. The completion of the azimuth is a full 360 degreesIn simplest terms, it reads like a compass for land navigation purposes.

The definition of Elevation is the measurement of an angle in respect to the Earth ellipsoid and the tangent plane of the Earth based on the true or astronomical horizon. The range is the measurement of distance between two objects (see Figure 14.9).

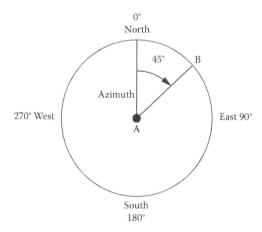

**FIGURE 14.8**
Defining an Azimuth angle.

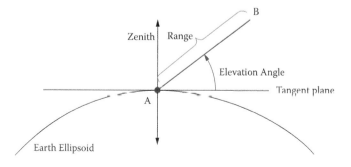

**FIGURE 14.9**
Defining an Elevation angle.

There are many things that can obscure intervisibility. Objects such as vehicles, buildings, mountains, and even the curvature of the Earth can impede a positive Access computation. Access uses calculations that consider what one can view in a straight line. Line of sight (LOS) is basically what is viewed within a straight line. LOS is a constraint option and is on by default. LOS many be turned off to allow visualization through an object within the calculations that allow valid Access intervisibility. Other impedances to intervisibility can be set within the Constraints section of an object included in the calculations.

As a special note regarding Access constraints and line-of-sight (LOS), in version 9 of the software, when using the LOS constraint with ground-based objects and using the Access tool to compute to a planet, the LOS constraint does not limit visibility with the planet. In earlier versions of the software, if Access was run between a ground-based object and a planet, LOS would need to be turned off.

## Customization of Data Providers

The customization of the data providers used within the Access interval is possible for the STK Tool and for each STK Object used within the calculations. You may customize what form of Vector Geometry you would like to report on, a special category on the Link Budget, or specific filtering results. Data providers are the low-level attributes and features that make up the whole STK Object or STK Tool. Gaining access to the data providers through the customization of the graphs and reports, the Calculation or the Time Tool allows you to specifically narrow down and refine your analysis. Since this is primarily used when customizing your output data, data providers are more directly discussed within the STK Output section of this book (Chapter 20).

## Deck Access

Deck Access is a form of Access. It evaluates the intervisibility from a single STK Object against a Deck or an array of STK Objects. However, this tool is different. It computes from the originating object using the spatial position and the time interval of the initial object. It then analyzes the position for intervisibility against a database of objects to find what object might be in the given look-angle during the allotted time. File types it can look for are satellites, stars, facilities and targets, star collections, and area targets.

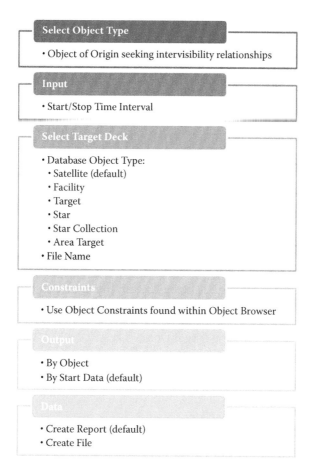

**FIGURE 14.10**
Deck Access options.

- *Select Object Type:* The Select Object Type option is the selection of the originating STK Object.
- *Input:* Input is the time interval criteria used to narrow down the possible targeted deck objects that could be available for intervisibility to the originating object. The default time interval is the Scenario time.
- *Select Target Deck:* This is the object type and database file selection area for your targeted deck object. Here you can select what type of object you would like to define and select to find intervisibility to the object of origin. The objects that can be used in the target deck are Satellites (default), Facilities, Targets, Stars, Star Collections, or Area Targets.

- *Constraints:* The Constraints option allows you to select a constraint (an instance of the object class) from an object within your Object browser to be used within your target deck object.
- *Output:* The Output option allows you to select how you would like to view the report or file, either by Object or by the default option, Start Date.
- *Data:* The Data output for Deck Access Create Report (default) or to create an output file to be used in another format.

To understand the difference between Access and Deck Access, you should compare the questions that are asked and the output results that each tool provides. For instance, for the STK Access tool, the questions Access answers best are "When can I see STK Object A, B,..., or *n*?" "How long can we see these objects?" "What elevation angle are these objects at?" The output results return all of these specific answers of intervisibility in the form of a static or dynamic map, graph, or report. However, with Deck Access, you are asking the question, "What STK Object from my provided catalog is in my field of view during a specific time interval?" The return from this would be a report of any available object that is listed in the database that matches the time interval selected for the Target Deck.

# 15

## Chains

---

### Objectives of This Chapter

- Define a Chain
- Chain Logic
- Chain Options

---

### Chains

Chains are combined segments of Access to create multihop relationships and include logical operations within the calculations. The Chain STK Tool creates a consecutive intervisibility relationship between two or more objects. The STK Objects that are evaluated may have as simple as a 1 to 1 (1:1) relationship or even a Many-to-Many-to-Many ... (M:M:M...M) for an entity relationship. Chains use a special STK Object, the Constellation. This leverages the array-holding capabilities found within the array and allows for robust pair-wise comparison between each intervisibility segment. As each leg is evaluated for valid Access, it then moves to the next leg until all legs show valid Accesses, including the STK Object attributes and constraints.

Chain relationships define the distance, duration, and elevation angle for each Access interval quantitatively. Much like the Access tool, Chains (see Figure 15.2) also evaluate signal relationships qualitatively. This means that the forms of ontology consider the relationship as a part where each segment is evaluated from the definitions of each STK Object to the definition of the Access or Signal. In addition, the ontological evaluation of the relationship as a whole Chain is evaluated, considering that in order for the Chain to be valid, every leg, or strand, within the multihop evaluation requires that each part has a true intervisibility relationship. The strands of Access are then overlapped and validated for complete Access from the beginning STK Object node of the Chain to the terminal STK Object node.

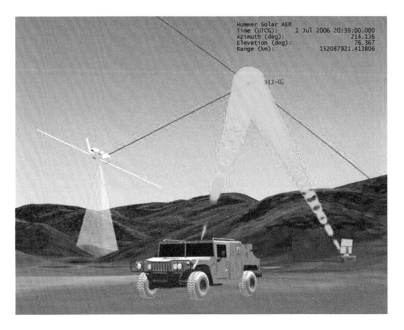

**FIGURE 15.1**
Chain Access from ground objects to satellites and aircraft.

**Chain Access Logic Example**

When:
- **Object A** has a Valid Access with **Constellation Object B** and
- **Constellation Object** B has a Valid Access with **Constellation Object C** and
- **Constellation Object** C has a Valid Access with **Object D**·

Then, we have a complete chain access.

Defined Domain: STK
Global Properties

**FIGURE 15.2**
Chain Access logic example.

## Chain Logic

Chains extend the pair-wise event detection found within the Access tool and allow for multihop linkage with a larger set of entities. This list of objects within a Chain is considered ordinal from the beginning of the set within the Chain to the end. In addition, the use of the Constellation object allows for a single node on the data relationship to point to several objects that have been grouped together within an array. This capability leverages the computational complexity to consider multiple forms of analysis.

In order for Chain Access to be calculated, there must be intervisibility along the signal path of all of the entities in the chain. Similar to the Access tool, by default, Light Time Delay is considered in the calculations. For a simple Chain event, with only two entities,

entity A AND entity B where A AND B = True

(Figure 15.3). This is using the exact same logic as the simple Access tool in a 1:1 relationship.

The need for a required AND between each entity would still hold true as the multihop sequence is further developed. For instance, the multihop logic would look like this:

entities A AND B AND C AND D = True

to have a complete chain access event.

The next form of Chain Access uses an object called a Constellation that has assigned other available objects to be considered within the calculations of the event. The Constellation reads as an array of possible objects to be read within the event. The default parameter between each item within the

Access

When **Object A** has a Access with **Object B:**

This is a 1·1 Relationship

Defined Domain: STK
Global Properties

**FIGURE 15.3**
Simple Chain Access.

**Chain Access Logic Example**

When:
• **Object A** has a Valid Access with **Object B** AND
• **Object B** has a Valid Access with **Object C** AND
• **Object C** has a Valid Access with **Object D:**

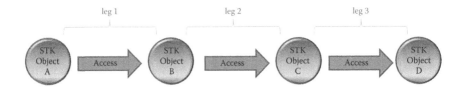

Then, we have a complete chain access.

Defined Domain: STK
Global Properties

**FIGURE 15.4**
Chain Access logic strand.

array has an "either/or" logic value. This means that in Figure 15.5, the first entity here represents an object and the second object is a Constellation that represents many objects. This is considered a many-to-one relationship. This Constellation has assigned several ($n$ value) items within the array so that

A AND B1 either/or A AND B2 either/or … A AND BN = True.

The logic within this simple Chain is the same logic that is represented in the simple Access event with default parameters shown in the Access tool chapter (Figure 15.6).

Chain Access may also combine the multihop structure and leverage the use of the Constellation object. The logic of the multihop will still compute similar to the previous multihop where

entities A AND B AND C AND D = True.

Here, we have defined Constellations B and C differently.

The arrays they hold are probably different forms of STK Objects. If we look at B, the objects could be a set of repeater ground antennas or they could be a set of facilities. For Constellation C, we could be considering an array of GPS satellites to be assigned within the constellation. In Constellation B, the array has maintained the default constraint of "either/or" logic within the definition. In contrast, Constellation C has a user-defined constraint of "At Least $n$" where

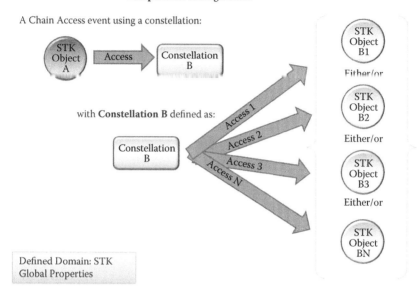

**FIGURE 15.5**
Object to Constellation logic.

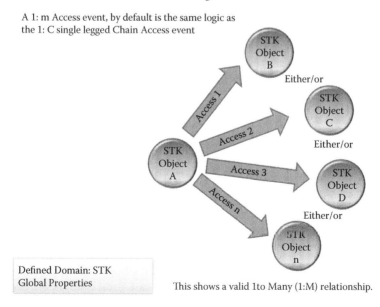

**FIGURE 15.6**
Access logic.

"$n = 4$." This means that in order for the Chain to be a valid Access Chain, a minimum of four STK Objects with four branch segments within a Constellation must be connected (see Figure 15.7). This modifies the basic computation

[(A AND B1) either/or (A AND B2) either/or (A AND B$n$)] **AND**
[(C AND D1) **and** (C AND D2) **and** (C AND D3) **and** (C AND D4)
or (C AND D$n$)] **AND** E = True.

## Basic

The Chains/Basic page is where the Chain Access Tool is initially defined. In the Available Objects list, the objects as displayed within the Object browser are redisplayed within the tool (see Figure 15.8). You may select any object from the list and assign it to the Assigned Objects list. Child and parent objects may be brought over for a Chain Access; however, they are computed individually for positional values. In addition, the order the objects are listed within the Assigned Objects list is the order the STK Engine computes the Chain Access. The order of computation begins at the top of the Assigned Object list and works systematically down each object. I recommend all users to make sure the order is correct by reviewing the sketch drawn when the scenario was originally defined. This is especially important if calculating a communications multihop.

Basic/Advanced Chain Access options allow you to modify the time period used for the Access calculations, which can be based on the Object (default), the Scenario, or a specified time period. At times, you may want to

### Chain Access Logic Example

When:
- **Object A** has a Valid Access with **Constellation Object B** AND
- **Constellation Object B** has a Valid Access with **Object C** AND
- **Object C** has a Valid Access with **Constellation Object D** (constraints "at least" 4):

Then, we have a complete chain access.
Note: This is a 1:M:1:M relationship.

Defined Domain: STK
Global Properties

**FIGURE 15.7**
Chain logic with Constellations.

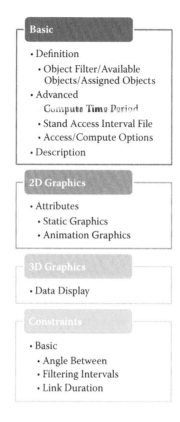

**FIGURE 15.8**
Chains Options Panel.

specifically identify the time intervals for the availability of an STK Object as a constraint. To do this, you create a Strand Access Interval file (*.int). This allows customization of the start and stop time intervals. Formatting the interval file needs to be consistent with the STK procedures found on the Interval List file format (*.int).

## 2D Graphics

The 2D Graphics page modifies the visualization of the Chain Access. Most of these options are inherited by the 3D Graphics window unless you choose not to inherit it. You have options to change the static and animation line width and color.

## 3D Graphics

Some of the dynamic data display options are controlled within the 3D Graphics page. This option allows you to display data provider information dynamically, as the Chain Access changes over time. Some of the options you can display relate to communications signal quality, such as Bent Pipe or Digital Repeater Link analysis. If more data provider information options are wanted to display within the 3D Graphic window, it is easier to go through the Report and Graph Manager. We discuss the Report and Graph Manager in the Reports and Graphs section of Chapter 20.

### Constraints

Chain constraints concentrate on how each relationship between the objects during the pair-wise analysis is evaluated. Constraint options for Chain Access are computations of the true position of the STK Object and the angle between these objects, the use of a custom interval file, or even the time duration of the Access interval. These constraints are in addition to the constraints specified on each STK Object.

# 16

## Coverage

### Objectives of This Chapter:

- Coverage Tools
- Figure of Merit
- Object Coverage and the Figure of Merit

### Overview of Coverage

Coverage provides both qualitative and quantitative time-dynamic analysis constrained to an Area Target, STK Object path, or Attitude Sphere. It analyzes by pair-wise comparison the intervisibility from an area of interest to a group of assets usually held within an STK Constellation Object. There are three separate forms of Coverage tools in version 9.*n*. First, the Coverage tool is found in the STK Objects/New Objects wizard. Secondly, the STK Object Coverage tool is found by right-clicking on the STK Object within the Object browser. Lastly, the Attitude Sphere Coverage tool is also found in the STK Objects/New Objects wizard, but in the Attached Objects section where the child objects are. These tools are primarily analytical and become visual with the statistical computation when combined with the Figure of Merit child tool. Coverage runs the Access algorithm much like the Access and Chain tools. Access determines intervisibility from the grid points to the assets it is evaluating.

Coverage properties pages require semantically developed attributes for the STK Objects to be meaningful. The Coverage tool definition is where the ontological relationship is formed. In order to visualize the Coverage definition, you need to define an evaluation based on statistical analysis from within the Figure of Merit. The Figure of Merit (FOM) is attached to the Coverage definition and inherits the Access properties to allow the statistical and visual response. These tools evaluate static or animated outputs. The static output

Three forms of the Coverage Tool

**FIGURE 16.1**
Three primary forms of the Coverage tool .

is usually a summation of the calculations over the specified time period by default. Animated graphics can show how the results are accumulated or what the evaluation results look like at a specific moment in time.

Coverage tools are used to answer such questions as how many assets can cover an area during a given length of time. They can also help you to understand if you have intervisibility gaps, run parametric studies, or be able to check for Dilution of Precision measurement or scalar calculations. Signal evaluations also can be evaluated for linkage quality.

## Coverage Tools

In general, the Coverage tool is an Access computational tool designed to work over a set area, object path, or attitude sphere. The analytical definition of the event detection is calculated based on the intersection points of a grid that is user defined and assigned to a polygon-shaped region of interest named an Area Target, a sphere, or longitudinal- or latitudinal-bounded areas. This Coverage over an Area Target bounded region can be contained within a sensor swath, set static on a central body, or draped over terrain. Some forms of Coverage may also use a polyline that has been propagated from a vehicle over an interval of time. Event-detection options allow for finding precise event times, time convergence, or value tolerance with step-size control.

The properties pages for Coverage define the types of grids to be used in the area of interest, the point granularity for the intersection points, and the location of the points. The grid points can be identified by Latitude, Longitude, Global, or Custom Boundary, or by Region. The grid points may either be computed by Point Granularity or a custom Point File that is imported.

With Coverage, the entity relationship is either a many-to-many or a many-to-one relationship. The origin point for calculating Coverage is based on the intersection points of the grid. The grid definition, by default, is set by points based on granularity of area from the latitude and longitude from the equator. It may also be defined by equidistance points or user-defined latitude/longitude points of interest based off location of the grid (G), independent of the higher or lower latitudes. The grid points do a pair-wise comparison with the STK Objects either individually or within a Constellation (C); see Figure 16.3.

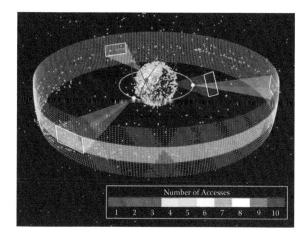

**FIGURE 16.2**
Coverage Access of a satellite .

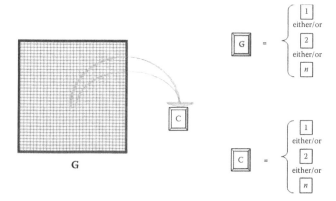

**FIGURE 16.3**
M:M coverage.

## STK Object Coverage

The STK Object and object path use the STK Object Coverage tool much like the standard Coverage tool. The path is modeled over time and visualizes a series of movement points, also known as waypoints. In the STK 3D Graphic window, this path forms a polyline. The STK Object Coverage tool is found by right-clicking the STK Object in the Object browser or selecting it from the STK Tools task icon menu. This polyline holds parallel points for pairwise comparison to the assets defined within the Object Coverage tool. One difference with the Object Coverage tool is that it has the Figure of Merit built into the tool interface, in contrast to the standard Coverage tool and the Attitude Coverage tool, which requires a child Figure of Merit attached to it.

### Attitude Coverage Tool

The Attitude Coverage tool is a specialized tool used if the STK Object has defined an Attitude Sphere within the STK Object/3D Graphic/Attitude Sphere properties page. The Attitude Coverage tool is an attached child to the STK Object and is found within the STK Attached Objects section of the software. The Attitude Coverage tool also requires a Figure of Merit, called an Attitude Figure of Merit (AFOM), to define the statistical analysis and allow visualization. This Attitude Figure of Merit is attached to the Attitude Coverage Tool, which makes this the child-of-a-child relationship.

Pair-wise comparison calculations for the AFOM are slightly different from the other coverage tools. The AFOM is unique due to the spherical nature of the Attitude, which is reflective of how the attitude coverage drapes over the STK Object. The calculation points are not based on grid point intersection but on the centrality of the STK Object and the area around the object. This relationship considers the STK Object with the Attitude Sphere and the evaluation from that to the Assets that are relevant to the ontology.

### Figure of Merit

The Figure of Merit (FOM) tool is an STK child to the Coverage tool. Coverage is not visual within the graphics of the STK GUI until it is coupled with the

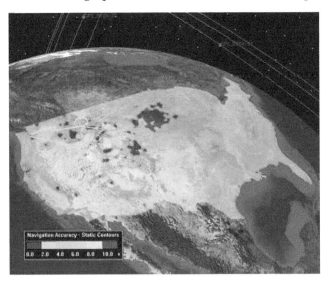

**FIGURE 16.4**
Navigation accuracy coverage event.

Figure of Merit as a parent–child relationship. The Figure of Merit acts as a child tool that inherits the relationship definition from the Coverage tool. It measures the efficiency of the "coverage" of the objects and returns a statistical response to both a graphical output and reports after it is computed through the STK Engine. The FOM adds to the algorithm the statistical evaluation of the relationship formed using the Coverage tool and defined by the attributes within the STK Objects.

There are many different measurements of efficiency that are used with the FOM tool. Most of the measurements are available for all three forms of the FOMs with the exception of the Dilution of Precision and Navigation Accuracy. These two measurements are not available within the Attitude Figure of Merit (AFOM).

- *Access Constraint:* This simply reports the average, maximum, minimum, percent above, percent below, and sum data results from the STK Objects as the assets have a valid relationship with the grid points defined within the Coverage tools. There is a finite list of possible constraints from the objects to choose from. This means that while not all of the possible constraints held within an object may be used, most of the basic and signal constraints are included in this list.

- *Access Duration:* Access Duration defines the time interval each asset has intervisibility.

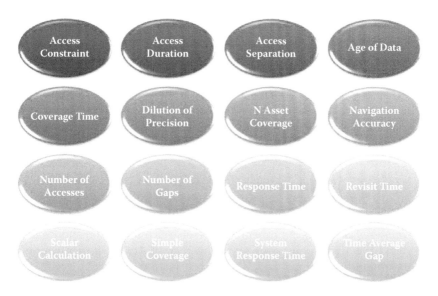

**FIGURE 16.5**
Access constraint attributes.

- *Access Separation:* Access Separation defines the time periods a valid Access is found. This produces a Boolean response for output that will give either a positive or negative response if Access is met for that area or not for the given time interval.

- *Age of Data:* The Age of Data defines the time an Access interval has stopped up to the current time of the analysis.

- *Coverage Time:* Efficiency measure of the duration there is coverage over the selected region of interest is both dynamic and static. This produces results for minimum/maximum times per day, minimum/maximum percent per day, and the percentage of time duration. Total values of valid Accesses are also evaluated.

- *Dilution of Precision:* The Dilution of Precision (DoP) measures the quality of intervisibility of the assets based on the geometric position of the assets. This is interpreted as a ratio of positional error to range error. The focus of this ratio is on the horizontal and vertical spread of the satellites as the assets have valid Access to a grid point or point. The wider spread of geometry of satellite, the lower the error ratio is, which translates to a lesser value of DoP. This FOM allows for the use of specific constraints, such as a minimum number of assets required to have Access during the same interval of time. DoP methods are based on a focus of the geometry as it is defined. The methods are geometric, positional, horizontal, vertical, eastern, northern, or time Dilution of Precision.

- N *Asset Coverage:* This evaluation gives the numerical values of assets that have valid Access and the times their accesses are valid. FOM Tool constraints are based on the evaluation.

- *Navigation Accuracy:* The Navigation Accuracy FOM considers the uncertainty of the navigation based on the number of satellites that have coincidental Access to a grid point. Since this is focused on grid capabilities, it is not available for Attitude FOM systems. Methods used in the calculations are based on geometry, including horizontal, vertical, eastern, and northern accuracy values. It also includes the time accuracy.

- *Number of Accesses:* This gives the calculation of independent or unique STK Objects as they have Access to the defined Coverage area. This also reports the time intervals these unique accesses are valid.

- *Number of Gaps:* The Number of Gaps returns a Boolean response for each time a grid point is in a coverage gap of Access. It then counts the invalid responses and returns an average number, a span of time, the number of gaps per day, a maximum or minimum number of gaps per day, or a total number of gaps during the full time interval of coverage.

- *Response Time:* Response Time measures the gaps or time between the valid Access request and the time the Access is considered valid.

- *Revisit Time:* This measurement focuses on the gap time interval duration of the current gap.

- *Scalar Calculation:* A Scalar Calculation considers the time-varying magnitude of angles, numerical derivatives of other scalars, integrals, or quadratic scalar combinations during the intervisibility. This is also used with the Calculation and Time tools and is new to STK version 10.

- *Simple Coverage:* Simple Coverage is a Boolean response regarding whether you have a valid or invalid Access calculation from the grid point to an asset.

- *System Response Time:* System Response Time is much like the standard Response Time, but instead the concentration is on a time interval from when an Access request is made and the time the full asset group is able to respond.

- *Time Average Gap:* This is a simple measurement of the average gap duration found through the time span of the coverage analysis.

# 17

## Communications

### Signal Communication Basics

Signal Communication uses the Access algorithm as the underlying determinant for intervisibility between objects during a pair-wise comparison. Signal linkage is evaluated qualitatively by the primary STK Tools: Access, Chains, Coverage, or CommSystems. Basic signal linkage is calculated by specialized STK Communication child object types, such as receivers, transmitters, antennas, and radars used in the evaluation. A signal evaluation cannot be determined from objects that do not have signal format data providers built within the STK Object. Therefore, objects such as a facility, aircraft, satellite, or sensor are not able to derive a signal evaluation. The use of STK Communication child objects and the ontological relationships based on the STK Communication tools allow for signal quality evaluations, interference, and jamming analysis.

All STK Communication tools evaluate signal by using the Access algorithm first. If there is intervisibility between STK Communications child objects while using Access or Chain Access, then qualitative analysis is computed. Intervisibility is evaluated in the exact same manner as discussed in the Access and Chains chapters. With simple Access, there are two forms of data relationships: one-to-one and one-to-many. The Chain Access handles pair-wise comparisons with data relationships as simple as a one-to-one relationship and as complex as a many-to-many-to-many... to-many relationship. With signal evaluations, multihop communications can not only be computed directionally in a bent-pipe but also other in forms of linkage analysis per each leg.

**Chain Access**

**Object A** has a Chain Access with **Object B** when:
**Object A AND Object B establish intervisibility, this equals a valid access.**

This is a 1:1 Relationship

Defined Domain: STK
Global Properties

**FIGURE 17.1**
Chain Access.

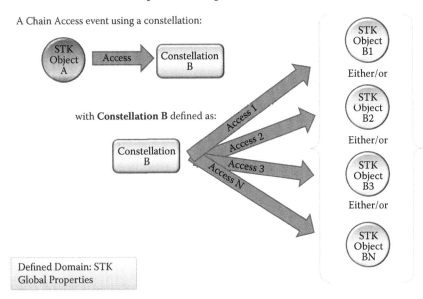

**Object to Constellation
Comparisons during Access**

A Chain Access event using a constellation:

with **Constellation B** defined as:

Defined Domain: STK
Global Properties

**FIGURE 17.2**
1:M Chain Access with a Constellation.

**Chain Access Logic Example**

When:
- **Object A** has a Valid Access with **Constellation Object B** AND
- **Constellation Object** B has a Valid Access with **Object C** AND
- **Object** C has a Valid Access with **Constellation Object D** (constraints "at least" 4):

Then, we have a complete chain access.
Note: This is a 1:M:1:M relationship.

Defined Domain: STK
Global Properties

**FIGURE 17.3**
1:M:1:M:1 Chain Multihop Access.

The tools used to compute communications evaluate different ontological relationships between the defined STK Communication child objects. For instance, the STK Access tool considers the signal quality and can derive a link budget analysis between the receiver child tool and the transmitter child tool. Duration, azimuth, and elevation angles are also calculated for each Access event. Chains evaluate signal between the signal quality of receivers, transmitters, and retransmitters also. However, just like the Chain evaluations is required when not using signal, every leg within the multihop to have a complete and valid intervisibility to ensure the capability of signal evaluation. Multihop communications also leverage robust pair-wise analysis using Constellations. Since up through version 9.0 the transmitter versus the receiver, calculations are based on transmitted, we have included a flow diagram in Appendix C.

The components used for the attributable definition of each Communications Object within a signal evaluation are the Communication components as discussed in Chapter 11. These components have assigned data providers that allow for a variety of outputs after the ontological relationship is established using the STK Tools and computed by the STK Engine. The components allow for more detailed STK Object development and the customization of gains, losses, or constraints scripting using MATLAB.

Because STK is a Spatial Temporal Information System, cartographic visualization of communications is built in by design. First, the STK Objects are defined, including the attributes within the 2D and 3D Graphic window properties pages, and then the ontological relationship of communication quality is selected. The STK Engine then computes the defined relationships and

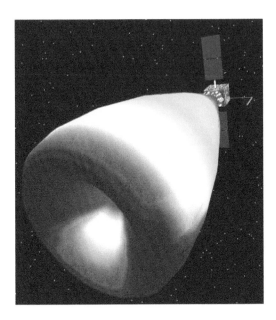

**FIGURE 17.4**
Visualization of antenna strength.

gives output instructions both visually and dynamically within the graphic interface. In addition, output may be in the forms of dynamic displays, graphs, and reports. Our focus for this chapter is to understand the abundance of qualitative ontologies that Access, Chains, Coverage, and CommSystems provide. This is not a comprehensive evaluation of STK Communication but an introduction. Comprehensive coverage of STK Communications is worthy of a book by itself.

## Access and the Link Budget

Communications analysis using Access allows you to evaluate link budgets and individualized signal quality for individual legs. Because the tool is using the fundamental Access STK Tool, it computes the event detection's time, time convergences, elevation angle, azimuth angle, and range of the event. The use of light time delay is a default in Access calculations. In addition to this, the Access tools calculate the gains and losses of a signal during communication. This is also known as a link budget. The final assessment of the signal is a defined number of bit errors over the total number of bits during transfer, or the bit error rate (BER). Reports for a link budget are discussed in the Output section in Communications output.

**FIGURE 17.5**
Link Budget definitions.

Figure 17.5 uses the standard help file definitions as they apply to the basic link budget. The simplified version of this link budget is very limited in results, but is sufficient for us to begin understanding how STK handles the computation. The detailed link budget has much more information.

A simplified link budget is computed from the Transmitter child object to the Receiver child object. The fundamental parameters of the transmission are configured by the attributes and component settings within the object. The linkage relationship through Access evaluates the quality of the signal setting from the transmitted node to the received node.

The basic default settings on the objects consider the Transmitter object output factors of the effective isotropic radiated power (EIRP) in the link direction. The Help files show how this is computed within STK. The EIRP value is the power of the transmitter (in dB), plus the antenna gain (in dB), plus any filter and post-transmit gains/losses (in dB). In the simple and medium Transmitters, the default antenna is preselected for you. However, the Complex Transmitter or the Multibeam Transmitter components allow you to define attributes for a specific antenna type from the STK Communication Components list. The default antenna is an omnidirectional point antenna without any nulls in the pattern. Although the EIRP is user defined in all antenna components, the gain on any antenna is not user defined for the simple antenna component. Unlike the Simple transmitter, the antenna gain is user defined in the Medium, Complex, and Multibeam Transmitter types.

The EIRP algorithm from the STK technical documentation is:

$$EIRP_i = P_t + G_{ant} - L_{filt} + L_{post}$$

where
$P_t$ is the transmitter power
$G_{ant}$ is the transmitter antenna gain
$L_{filt}$ is the filter loss
$L_{post}$ is the post-transmit gains/losses

Receiver output computations, as found in the STK Help files, evaluate the received Doppler shift and the received isotropic power (RIP) with the flux density calculations as the linkages are established. These computations are found within the STK technical notes on how the signal is derived.

Received Doppler Shifted frequency calculations are made using:

$$f_{Rec} = f_{In}\sqrt{\frac{c - \dot{r}}{c + \dot{r}}}$$

where
$f_{In}$ is the transmitted frequency
$c$ is the speed of light
$\dot{r}$ is the relative velocity between the transmitter and receiver

The received isotropic power (RIP) calculations use:

$$RIP = EIRP + L_{prop}$$

$$\Phi = RIP + \log\left(4\pi\frac{f^2}{c^2}\right)$$

where
$EIRP$ is the transmitter effective isotropic radiated power
$c$ is the speed of light
$L_{prop}$ is the total propagation losses between the transmitter and receiver
$f$ is the received frequency

Flux density calculations per STK technical notes use:

$$\Phi = EIRP \times L_{prop} \times 4\pi\frac{f^2}{c^2}$$

where
$EIRP$ is the effective isotropic radiated power in the link direction
$L_{prop}$ is the channel propagation loss

*f* is the received frequency

*c* is the speed of light

The bit error rate (BER) is computed by the number of bits in error divided by the number of bits in the stream. STK handles the calculation to simulate live bit errors with the use of the signal-to-noise ratio (Eb/No) and a lookup table from a correlated coded modulator file (*.mod) found in the install directory :...\\STKData\Comm\SrcModFiles. The results are interpolated from a given bit energy level. If the energy level is smaller than the first number in the table, then the BER for this value in the table is used. However, if the bit energy level is larger than the last number in the table, than the number indicates there are no errors and returns a value of $1.0^{-30}$. With BERs, the lesser the value, the better the signal quality.

In the Reports and Graph section, there is a more detailed link budget that includes all of the simplified link budget information and additional information. This information is described within the STK Help files, as shown in Figure 17.6.

## Chains: Single Hop and Multihop

Chains allow for both single-hop and multihop signal evaluations on STK Communication child objects. Identifying quality of signal using the STK Communications package allows for complex analysis using the STK Constellation Object and the ability to evaluate multiple levels of pairwise comparisons during one analysis. With proper setup of both the STK Objects and the STK Tools, the results can show all possible signal links for multiple paths.

Although the algorithm is established using the STK Objects and the Chain tool, the calculations for signal evaluations are handled within the Report and Graph Manager data providers and then computed using the STK Engine. Chains allow for all link budget analysis, as shown by the simple Access Link Budget and the Detailed Link Budget. This means that all basic link budget analyses that are shown in the Access Link Budget are reported in Chains. In addition, data providers allow for bent pipe and retransmit signal evaluations, and the ability to evaluate total linkages from multiple legs.

Additional fields available for Chains are shown in Figure 17.7.

Bent pipe evaluations are used within the Chain tool for both digital and analog signal evaluations.

**FIGURE 17.6**
Advanced Link Budget definitions.

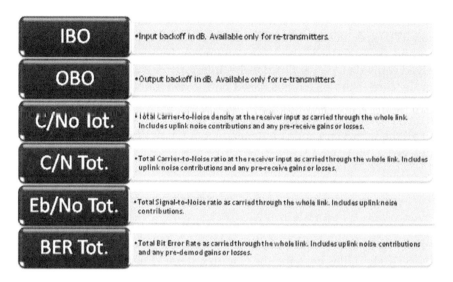

| IBO | •Input backoff in dB. Available only for re-transmitters. |
| OBO | •Output backoff in dB. Available only for re-transmitters. |
| C/No Tot. | •Total Carrier-to-Noise density at the receiver input as carried through the whole link. Includes uplink noise contributions and any pre-receive gains or losses. |
| C/N Tot. | •Total Carrier-to-Noise ratio at the receiver input as carried through the whole link. Includes uplink noise contributions and any pre-receive gains or losses. |
| Eb/No Tot. | •Total Signal-to-Noise ratio as carried through the whole link. Includes uplink noise contributions. |
| BER Tot. | •Total Bit Error Rate as carried through the whole link. Includes uplink noise contributions and any pre-demod gains or losses. |

**FIGURE 17.7**
Additional data from multihop communications.

## Coverage

Coverage, with the use of the Communication systems, allows for analytical computations along with visual and reporting output of the signal strength over a region of interest or vehicle path. The Coverage tools combined with the correlating Figures of Merit allow you to evaluate specific signal quality and efficiencies over a region of interest defined within the Coverage tool and the specific STK Communication Objects. Jamming contours can be displayed by using the Coverage tool along with the CommSystems with your Communication objects. This allows you to evaluate both jamming and interference. For more advanced features, you can evaluate TIREM or Urban Propagation, and it is interoperable with QualNet and GIS.

### CommSystems

The CommSystems tool allows you to identify possible sources of interference from vehicles and low to medium Earth satellite communication signals that have been established in your scenario. Linkage within the CommSystems tool utilizes signal analysis of Constellations of receivers and transmitters against the possible signal interference source that is also placed within a Constellation. Interference source or reference bandwidth may either be computed for a specific time or for an interval of time.

Interference methods work within the International Telecommunications Union (ITU) specifications. The ITU governs the global use of the radio spectrum and the standards used for equipment. Interference focuses on the received equivalent power flux density. Interference is defined as the overlap from the carrier-to-noise ratios based on the denominator of the ratio from the receiver and transmitter bandwidth. The CommSystems' link information data provider reports the total power flux density (Power Flux Density) from all the interferers and is independent of the desired or intended transmitter, given by:

$$Pfd_{tot} = \sum_{i=1}^{n} Pfd_i$$

where $n$ is the number of interferers.

The STK Help files describe the CommSystems' Interference Information as calculated using the Power Flux Density per interferer, which is independent of the desired or intended transmitter. The following is the equation behind STK's Power Flux Density. It is important to remember that all units are in a linear scale.

$$Pfd = EIRP_i \times L_{pol} \times \frac{1}{4\pi \times D^2} \times \frac{G_{rcvrAnt}}{Gmax_{rcvrAnt}} \times \frac{B_{ref}}{B_i}$$

where

$EIRP_i$ is the effective isotropic radiated power of the interferer in the receiver's direction

$L_{pol}$ is the polarization loss of the interferer

$G_{rcvrAnt}$ is the antenna gain in the direction of the interferer

$G_{maxrcvrAnt}$ is the maximum antenna gain (bore sight gain)

$B_{ref}$ is a reference bandwidth

$B_i$ is the bandwidth of the interferer

$D$ is the distance between the interferer and the receiver

The CommSystem is a complex and dynamic STK Tool that establishes the ontological relationship to evaluate qualitative analysis of signal similarities, overlap, and differences. Because the analysis is qualitative, the composition and constitution of the signal becomes the formal analysis.

## Communication Extensions

### TIREM

Terrain Integrated Rough Earth Model (TIREM) is a physics-based modeling software. This software allows you to analyze median propagation loss from a range of 1 MHz to 40 GHz. It is used as a global add-on extension at the scenario level for STK. It is introduced to the scenario within the Scenario properties page under the RF page section. TIREM was created by Alion Science and Technology Corporation. (http://www.alionscience.com) based in Annapolis, Maryland. This software examines the propagation of land and sea paths for terrain types and elevation profiles to endpoints that are either at line of sight or beyond line of sight of a central body. Alion states that their software evaluates free-space spreading, reflection, diffraction, surface-wave, tropospheric-scatter, and atmospheric absorption. Because of the ability to use the profiles delivered by the Alion product, it leverages the physics-based modeling used within the STK software environment and the STK physics-based geometry engine.

For points within the line of sight, TIREM path loss considers frequencies to determine the correct model to use for each analysis of sea only, sea/land combinations, or land only. Sea evaluations evaluate the Smooth-Earth Loss model, along with sea and land using the weighting by Sea Distances model from 1 MHz to 10 GHz. Land line of sight (LOS) includes models for Smooth-Earth reflection loss from 1 to 16 MHz, Interpolation of Reflection Loss from 16 to 20 MHz, and more rigorous Reflection Loss from 20 MHz to 10 GHz. The Tropospheric Loss model relates these above models to the troposcatter, or the measurement of how the electromagnetic signal is reflected off the Earth's troposphere and defraction in a beyond-line-of-sight (BLOS) analysis. It models a function of the frequency. Both LOS and BLOS models consider atmospheric absorption as well.

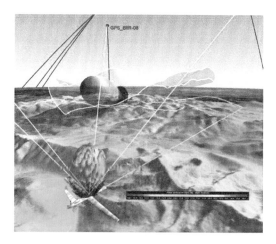

**FIGURE 17.8**
Multihop communications analysis.

### Urban Propagation

The Urban Propagation add-on model evaluates sight-specific propagation loss between two locations within an urban environment, such as an urban canyon. This Triple Path Geodesic model evaluates three dominant signal paths from the STK Transmitter to the STK Receiver object. Measurements are evaluated at the received signal for combined signal strength. This technology was developed Remcom, a business partner of AGI.

### QualNet

QualNet, created by Scalable Networks, computes network-centric systems in a real-time environment. Since STK models real-time, physics-based communications, the ability to allow network analysis becomes critical for battlespace management. STK Objects define the routes that leverage the physics-based modeling of movement and the dynamics with gravitational models within the STK environment. QualNet capitalizes on the interoperable use of the object-orientated environment to evaluate the wireless linkage and antenna output of the physical layer needed to understand wireless communications. Improvements on the interoperable use of STK are forthcoming and will only get better for more accurate and network analysis.

# 18

## Conjunction Analysis

### Objectives of This Chapter

- Define Conjunction Analysis
- Understand the Use of Prefilters
- Understand How to Perform Conjunction Analyses Using CAT
- Understand How CAT Calculates Probability of Collision
- Reviewing Some Best Practices

### Conjunction Analysis

STK provides a great ontological framework to not only see where individual satellites are, but also where they are in relation to each other. Displaying this information in a 3D window is a powerful way to visualize these relationships, as can be seen in Figure 18.1, which shows Earth-fixed orbits for all operational satellites in a geosynchronous orbit along with the dead satellites and other debris that threaten them.

This visualization can be downright scary, since this abstraction of reality makes it appear that many satellites are too close together to operate safely. But space is a big place and we need a tool that helps us efficiently sort through all of the data to see which conjunctions—or close approaches—we really should be concerned about.

That's exactly what STK's Conjunction Analysis Tool (CAT) does. Whether you are concerned with only a single satellite, a constellation of satellites, or the entire population of active satellites, CAT can quickly analyze one-on-many or many-on-many satellite scenarios to efficiently determine when satellites come within a user-specified distance of each other.

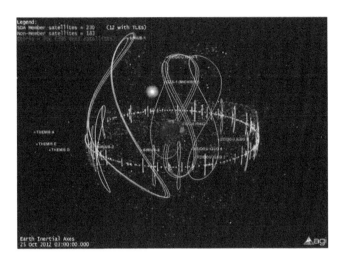

**FIGURE 18.1**
Geosynchronous satellites and debris.

## How Does CAT Work?

The underlying task is not particularly difficult. For each pair of objects (satellites or debris), CAT finds those times when the distance between those objects is less than a specific value. Doing this efficiently is where CAT shines. CAT interpolates the interobject range to quickly find minima (close approaches), avoiding the brute-force method of stepping along at very small time intervals.

But many pairs of orbits need not be examined at this level, since their basic characteristics prevent close approaches. CAT provides an easy way to prefilter the data to eliminate further analysis of orbit pairs that simply cannot produce a collision.

For example, if the apogee of one orbit is always less than the perigee of another, it is impossible for these objects to collide. Allow for a pad to compensate for the uncertainty associated with the individual orbits and you have an Apogee-Perigee filter—just one of four prefilters in CAT.

Even if apogee and perigee overlap, conjunctions can only occur at the node of the two orbits or along the line that defines the intersection of the two orbit planes. If the distance between the two orbits at these nodes is more than a specified pad, the two objects cannot collide. This prefilter is known as the Orbit-Path filter.

And, finally, if the objects aren't both near the same node at the same time, it is once again impossible for a collision to occur. This prefilter is known as the Time filter.

The final CAT prefilter is the Out-of-Date two-line element (TLE) filter, which discards TLE data over a certain age to avoid using data that will likely not be accurate enough to make a reliable prediction.

Applying these prefilters before performing the actual conjunction analysis, while requiring some additional computation, takes less time than having to examine the full set of possible conjunction combinations. Each of these prefilters can be individually selected and its pads set using either of two methods for accessing CAT.

## How to Use CAT

There are two ways to access CAT. The first is to right-click on the satellite of interest, select Satellite, and then Close Approach. The main panel allows you to set some basic conditions (e.g., time period, max range) and then perform the conjunction analysis computation. The Advanced button allows you to select the list of TLEs to screen against, set which prefilters to use and the associated pad values, and even have the conjunctional satellites automatically added to your STK scenario.

When you click Compute, the results of the conjunction analysis are displayed in the Compute Results window, as seen in this example:

- Close Approach initial Satellite Database search undertaken
- 15,010 candidate satellites found in Satellite Database
- 14,956 candidate satellites remain after date and propagation test
- 1,316 candidate satellites remain after apogee perigee filter
- 767 candidate satellites remain after path filter
- 322 candidate satellites remain after time filter
- 3 candidate satellites remain after range filter
- 3 satellites found to have access within the specified constraints
- Close Approach processing completed

As you might expect, Advanced CAT provides even more capabilities. Again, the prefilters and their pads are set on the Basic Advanced properties panel for the AdvCAT object. But Advanced CAT also has a couple of nice features to make the overall computation process even smoother. The Basic Advanced properties panel has an option to use a Related Objects file to avoid looking for conjunctions between objects that are known to be close—such as satellites in a geostationary cluster. Simply listing each cluster of objects as related not only removes reporting of close approaches, but it can significantly reduce computation time, since each close approach will not be prefiltered out and will require additional computation.

There is also an option to specify a file that contains the hard-body radii—the maximum width of a satellite—for use in probability-of-collision computations. This is very important, since modeled point masses would only collide with a direct collision, whereas in reality they can collide whenever the centers of mass pass within the sum of the two hard-body radii. For example, if two satellites—each of which has a hard-body radius of 4 m—pass within 8 m of each other, there is some likelihood (depending upon their attitudes) that they will collide. Or, more accurately, if they pass more than 8 m apart, they cannot collide.

Setting up and running Advanced CAT is done on the Basic Main properties panel. Here you can select one or more primary objects—typically the satellites you are concerned about—along with the secondary objects, those that may present a threat to the primary objects. Typically, the primary list is a subset of the secondary list since even operational satellites can present a threat to each other if not closely monitored. You can select any satellite objects saved as part of the scenario (regardless of the type of propagator used) or entire sets of TLE-based orbits (typically the main ones updated via the Data Update utility). Unfortunately, you cannot change the TLE reference directory (as you can with the Close Approach setup), so you need to place any custom lists in the same directory and they must have a .tce or .tle extension (.txt will not work).

If you are interested only in range-based calculations, simply check Use Range Measure, set the Threshold, and click Compute. Once that is done, you can right-click on the AdvCAT object to generate any CAT-related reports to obtain specific information on any close approaches. And, if you set the 3D Graphics attributes to show ellipsoids, you can see the results in the 3D window.

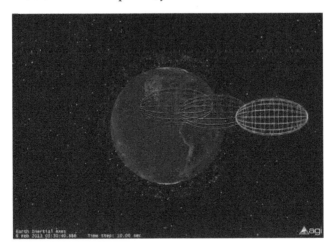

FIGURE 18.2
Advanced CAT results in 3D window.

## Calculating Probability of Collision

If you are interested in probability of collision, Advanced CAT is the tool you need. First, set the Threshold to zero and uncheck Use Range Measure. This setup will now calculate the distance between the covariance ellipsoids defined for your objects (the Tangential, CrossTrack, and Normal values set as the defaults when you added them) instead of the centers of mass.

When the ellipsoids touch, they essentially are at a distance corresponding to a combined covariance of 1 sigma and the ellipsoids will turn red in the 3D window. If you want conjunctions with 2- or 3-sigma levels, set the Ellipsoid Scaling Factor on the Basic Advanced properties panel to 2 or 3, respectively.

Note that the true or maximum collision probability (both are calculated) depends not only on the combined covariances but also on the combined hard-body radii. If you do not specify the hard-body radii, STK uses a default value of 1 meter, so a combined 1-sigma close approach may actually have a maximum probability of collision of $1^{-5}$ or less. The difference is that the combined 1-sigma ellipsoid is the probability that the two satellites are both within that volume, not the probability that the two objects of specific sizes will collide when a specified distance apart.

Setting realistic covariance values can be difficult, unless you have that data from an independent orbit-determination solution. For this case, you would set the covariance in the Satellite object definition and select the Covariance type in AdvCAT. This case will provide the most realistic results.

Using a Fixed Covariance type is easiest (and is the default) but not very realistic, since real covariance ellipsoids grow and rotate over time. STK does provide an option to define covariance growth as a quadratic, but the effort involved still does not result in a realistic covariance that rotates.

The Orbit Class options should be used with caution, since even though it seems intuitive that satellites in similar orbits would have similar covariance characteristics, detailed analysis using the GPS constellation seems to refute this assumption (see http://celestrak.com/publications/AAS/07-127/). This result arises because even though the orbits may be similar, the observation process may be different (e.g., different observation geometry, lighting, location of observations in the orbit).

## Real-World Applications

As you might imagine, it can take some time and attention to detail to properly set up an STK Scenario to look for close approaches with your satellites of interest. Fortunately, STK has some very powerful capabilities using its Connect language to automate structured tasks like these using C#,

VBScript, JavaScript, and more to ensure consistent application and avoid errors. Although we don't cover Connect in this book, we can offer an example to pique your curiosity.

CelesTrak offers a service called SOCRATES (Satellite Orbital Conjunction Reports Assessing Threatening Encounters in Space) that uses Connect to automatically create a scenario to screen all satellites in Earth orbit against anything that comes within 5 km of them over a seven-day period. A recent run screened 3,237 satellites against 15,028 total objects (all of those satellites and all the debris that we had TLE data for). It ensured that the proper Related Objects and Hard-Body Radius files were used while making sure all the other configuration options were properly set.

It took 80 minutes (on a standard desktop PC) to find 17,602 times when some object came within 5 km of those 3,237 satellites over a one-week period. The results are uploaded in a .csv file to CelesTrak, where Perl and JavaScript are used to report the results. If you want to see the Top Ten conjunctions by range for the coming week, just go to http://celestrak.com/SOCRATES/ and click the link. Often these conjunctions are predicted to come within 100 meters.

But the real power of STK comes in if you click the Analysis button for a particular conjunction (which requires Microsoft Internet Explorer (MSIE), since STK will be launched using ActiveX). The page that comes up shows the TLEs used and allows you to launch STK with a single button click and then build the scenario for that specific conjunction with another button click. All the settings for that scenario are set automatically (be sure to check those out once the scenario is created). You may want to review "What You Need to Know to Run STK with SOCRATES" to ensure the security settings for MSIE are configured properly to support running ActiveX.

Within a matter of seconds, you can launch STK, build the scenario, and jump to the threshold start. Then, when you click Play, you can watch the conjunction from the vantage point of the satellite as the threat whizzes by. The advantage of this type of service-oriented architecture (SOA), where the server generates the results for all users and the individual clients can then interact with those results to perform additional analysis, is one of the most powerful capabilities of STK. And the standardized process flow speeds the analysis, standardizes the presentation of the results, and minimizes the chances of errors.

# SOCRATES

## Search Results

Search parameters:

- Name(s): Top 10 by Minimum Range
- Order by Minimum Range
- Return first 10 items

*Data current as of 2013 Feb 07 12:45 UTC*

Computation Interval: Start = 2013 Feb 07 12:00:00.000, Stop = 2013 Feb 14 12:00:00.000
Computation Threshold: 5.0 km
Considering: 3,237 Primaries, 15,028 Secondaries (17,602 Conjunctions)

*See notes at bottom of page for data field descriptions*

Bookmark this search (Top 10 by Minimum Range)

Sort results by Maximum Probability, Time In.

| Action | NORAD Catalog Number | Name | Days Since Epoch | Max Probability | Dilution Threshold (km) | Min Range (km) | Relative Velocity (km/sec) |
|---|---|---|---|---|---|---|---|
| | | | | Start (UTC) | TCA (UTC) | Stop (UTC) | |
| Analysis | 08799 | METEOR 1-24 [?] | 9.070 | 8.605E-02 | 0.010 | 0.038 | 14.278 |
| | 13718 | METEOR 2-9 [?] | 8.129 | 2013 Feb 14 09:04:55.969 | 2013 Feb 14 09:04:56.320 | 2013 Feb 14 09:04:56.670 | |
| Analysis | 27939 | MOZHAYETS 4 (RS-22) [+] | 6.710 | 6.374E-04 | 0.020 | 0.082 | 14.667 |
| | 35664 | COSMOS 2251 DEB [-] | 8.276 | 2013 Feb 13 19:02:16.584 | 2013 Feb 13 19:02:16.925 | 2013 Feb 13 19:02:17.266 | |
| Analysis | 06787 | OPS 8364 (DMSP 4) [?] | 6.890 | 1.823E-04 | 0.057 | 0.100 | 3.834 |
| | 36005 | COSMOS 2251 DEB [-] | 7.876 | 2013 Feb 13 09:05:22.516 | 2013 Feb 13 09:05:23.820 | 2013 Feb 13 09:05:25.124 | |
| Analysis | 16011 | COSMOS 1680 [?] | 7.713 | 9.781E-04 | 0.024 | 0.103 | 14.912 |
| | 12172 | DELTA 1 DEB [-] | 8.290 | 2013 Feb 14 07:28:30.534 | 2013 Feb 14 07:28:30.870 | 2013 Feb 14 07:28:31.205 | |
| Analysis | 00022 | EXPLORER 7 [?] | 4.754 | 1.417E-04 | 0.061 | 0.120 | 4.634 |
| | 26782 | TITAN 3C TRANSTAGE DEB [-] | 5.314 | 2013 Feb 11 07:23:10.532 | 2013 Feb 11 07:23:11.610 | 2013 Feb 11 07:23:12.689 | |

**FIGURE 18.3**
SOCRATES top ten results.

**FIGURE 18.4**
SOCRATES conjunction analysis window.

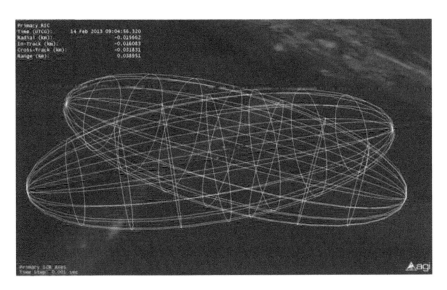

**FIGURE 18.5**
STK 3D view at time of closest approach.

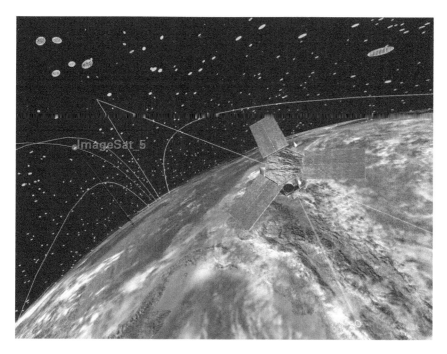

**FIGURE 18.6**
Satellite conjunction analysis.

# 19

## Nonontological Tools

### Objectives of This Chapter

- VGT
- Globe Manager
- Terrain Conversion
- Calculation and Time Tools (STK v.10)

### Overview of Nonontological Tools

STK uses tools to build relationships with the STK Objects. In addition, they also enhance the STK Objects' visual or analytical natures. The tools that build relationships are ontological tools, such as Access, Chains, Coverage, and CommSystems. Nonontological tools refine, either directly or indirectly, the STK Objects and relationships based on the STK Tool used. Some nonontological tools are the Vector Geometry, Globe Manager, and Terrain Conversion tools. In addition, the Calculation and Time tools, both new in STK version 10, leverage the STK Object understanding and the ease of customizing output and controlling the data. At this time, the Time and Calculation tools are also considered nonontological tools. This chapter is a high-level overview of the substance of these nonontological tools. All of these are taught within the STK comprehensive course.

### Vector Geometry Tool

The Vector Geometry Tool (VGT), also known as Analysis Workbench, enhances the STK Object attributes and constraints by applying or creating

```
RQ5_Hunter2 - Wellton LOS  at: T +01:06:24.847  (hms)
RQ5_Hunter2 - Yuma LOS  at: T +01:06:24.847  (hms)
RQ5_Hunter2 - Ajo LOS  at: T +01:06:24.847  (hms)
```

```
RQ5_Hunter2 Velocity Heading
Azimuth (deg):           66.63
Elevation (deg):          0.00
Horiz Rate (nm/hr): 180.00
Velocity (nm/hr):    180.00
```

**FIGURE 19.1**
Vector geometry and the STK Object.

additional geometric features. In reality, if you were to deconstruct the basic STK Object, you would find it begins with a point or a point mass. The STK Object, as discussed in Chapter 1, is a point, line, or polygon.

Each of these three basic geometric shapes begins with a point. The point mass is a point with vector capabilities. Points can also become lines and polygons. Points and point masses can have coordinate planes and systems. Angles can be created from the proximity or definition of the points, lines, vectors, planes, and other systems. In other words, the VGT allows you to add coordinate vector data to objects. This enhances calculations used for pointing, motion, and dynamics. The geometries become dynamic and ana-lytically strong. The VGT is used in customized vehicle attitudes, for orbits in custom planes, and to measure proximity and ranges between objects. In addition, the user can place the central point of an object to measure velocity, for example, or evaluate an angle of incidence for a rocket launch trajectory. Encapsulated geometries of the STK Object are defined by the VGT in reference to the STK Object's points, vectors, angles, coordinate systems, and planes. The VGT increases the computational and visual understanding of your STK Objects.

The VGT allows you to use either the established geometry component or a customized component in the form of angles, axes, coordinate systems, planes, points, or vectors. To derive geometric analysis, it uses the computational engine to coordinate the geometric operations within the coordinates from the STK Object or system selected. The computational engine passes calculated geometries and coordinates from one operation to another to verify the reference coordinates are compatible from one object to another.

When looking for compatibility, the computational engine considers the Access constraints, Sensor Pointing definitions, attitude/ephemeris, visualization, and the Report and Graph Manager of the STK Objects and Tools. This tool is used within the computational constraints of an object while Access evaluations are being considered. Indeed, it seems logical to say that the Vector Geometry Tool is perhaps one of the most fundamental and strongest nonontological tools found within STK. The Help files define the VGT customizations well. It is recommended that you use this chapter and the Help files to assist you in creating your own custom Vector Geometry for your objects.

## Angles

The Angles tool of the vectors allows you to define the distances between angles created from vector and plane components within the VGT. It can measure:

- Between two vectors
- Between two planes
- A dihedral angle between two vectors about an axis
- Shortest rotation between axes
- Between a vector and a plane

## Axes

Axes is a geometrical component created from custom vectors. It defines the reference point and form of rotational motion of an object. The Axes may be constrained or aligned using two vector references. It can denote an offset, B-Plane target body, libration point, trajectory reference, and spinning or fixed axes.

## Coordinate Systems

There are several types of Coordinate Systems defined and used within STK. The VGT allows you to use predefined coordinate systems and modify the central bodies system when needed. The forms of Coordinate Systems are defined in the technical notes of the VGT and are listed below as a quick reference. Your entire ontological analysis can change by varying what coordinate system you are applying. Since many of these are unique to the central body they are working with, it makes sense to spend some time in the technical notes when redefining the Central Bodies Coordinate System.

- ICRF
- J2000
- Inertial
- Fixed
- TrueOfDate
- TrueOfEpoch
- MeanOfDate
- MeanOfEpoch
- TEMEOfDate
- TEMEOfEpoch
- B1950
- AlignmentAtEpoch
- MeanEarth
- PrincipalAxes_403
- PrincipalAxes_421
- Fixed_IAU2003
- Fixed_NoLibration
- J2000_Ecliptic
- TrueEclipticOfDat

## Planes

Geometric Planes are defined through central bodies or STK Objects. There are four primary types of Planes customizations for the VGT: Normal, Quadrant, Trajectory, and Triad. You may create these planes on the default axes, or customize default offset rotational axes for more accuracy.

## Point

Points are a simple geometry either on an STK Object as a parent or in a stand-alone position within a plane or other system defined within a three-dimensional Euclidean space. As you understand how the software is really developed, you will realize the STK Objects are really point geometries. Some of these points are static, and some are point mass objects with the physics vector-based capabilities. This point geometry component is the basic building block for the STK Object structurally. Therefore, when we use the VGT to create points, we allow strong capabilities of being able to create points of reference and vector point mass objects using the VGT.

Points used within STK have various types based on whether they are fixed or kinematic points. Fixed points are specified within a designated coordinate system. Kinematic points are defined by the motion vectors. They are based on Central Body and Terrain references. In addition, points in motion may be derived from ephemeris files.

The position of the point of an object is based on the frame of the STK Object in reference to the central plane of the Central Body the object is within. The default fixed frame is the Earth's Central Body frame. Points are defined within the VGT as:

- Attached point to a 3D STK Object Model
- B-Plane point using a Direction, Target Body, or Type
- Central Body Grazing Point
- Covariance Grazing Point of a Central Body
- Glint point reflected from a central body
- The intersection point of a plane, vector, or origin point
- Libration point in reference to other primary or secondary bodies
- Point Fixed in a reference coordinate system
- Projection point in reference to a plane or other point
- Surface point as a secondary or sub-point to a central body

## Vectors

Vectors, by standard definition, have both magnitude and direction. Within the STK environment, this magnitude and direction are within a three-dimensional Euclidean space. Vectors can be attached to any point or used with angles, axes, other vectors, and planes.

## Custom Vector Geometries

The VGT is component based and may be customized. From the component list, there are three indicator colors: green, yellow, and orange. Both the green

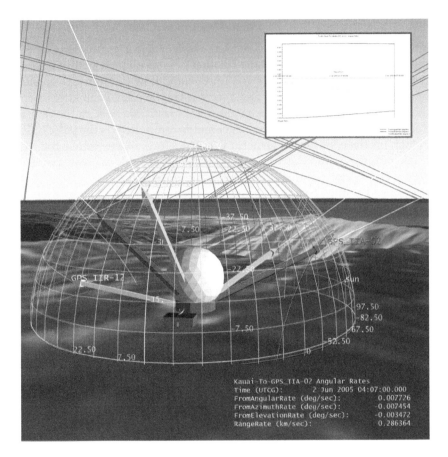

**FIGURE 19.2**
Vector geometry use.

and yellow components have the ability to be copied to use in a base template and modified. The orange components cannot be modified, copied, or deleted from the component list. There are two ways to customize vector components easily within the software: by visually selecting the components within the software environment and assigning new values and parent objects or by creating an Object Model file from the VGT Component Library Application Programming Interface (API) and importing it into the STK environment.

Within the STK VGT, an object is created by copying an old VGT and customizing it or by creating a new one. You may also customize or create a VGT using STK Object Model or the API. STK Object Model is the AGI term for a Component Object Model (COM), which is used with the Microsoft.NET technology. AGI has a full component library the Application Programming Interface (API) uses that includes the Vector Geometry tool. These tools may then be used within the customized environment using other parts of the AGI Components or be used as a plug-in with the STK environment.

## Globe Manager

The Globe Manager assists in the visual control of the 3D objects in the 3D window. It manages terrain data and imagery, as well as Microsoft's Bing maps. KML and .jpeg images work well within the environment as well as the AGI indigenous format of .pdtt and .pdttx file formats.

## Terrain Conversion

The Terrain Conversion tool is another nonontological tool. It converts the terrain data analytically so that it is formatted correctly to be visualized using the Globe Manager. Terrain conversion restructure the terrain files into .pdtt or .pdttx files to be used within the Globe Manager or scenario level of the software. File formats that can be converted for STK use are:

STK World Terrain (HDR)

ArcInfo Binary Grid (adf)

ArcInfo Binary Grid - MSL Vertical Datum (adf)

ArcInfo Grid Depth MSL

GEODAS Grid Data (g98)

GTOPO30 DEM (hdr)

MOLA Terrain (LBL)

MUSE Raster File (DTE)

NIMA/NGA DTED Level 0 (DT0)

NIMA/NGA DTED Level 1 (DT1)

NIMA/NGA DTED Level 2 (DT2)

NIMA/NGA DTED Level 3 (DT3)

NIMA/NGA DTED Level 4 (DT4)

NIMA/NGA DTED Level 5 (DT5)

NIMA/NGA Terrain Directory (DMED)

Tagged Image File Format (TIF)

Tagged Image File Format - MSL (TIF)

USGS Digital Elevation Model (DEM)

## Calculation and Time Tools

The Calculation tool and the Time tool are both new to version 10. The tools take advantage of the data providers used with the STK Objects and STK Tools to improve control over computations and reporting capabilities. If you go back and look at the general form of the STK Object and STK Tool, you will notice three primary categories of attributes found within the properties pages for each object: Geometric, Time, and Calculations. Both the STK Object and STK Tool are modified from their respective properties pages. These changes to the objects and tools are semantically detailing the components and subcomponents to automatically define the data providers. The time intervals for each object and tool autoreferences the scenario; calendar control features give you the ability to have more control over customization. In the future, a Volumetric tool is on the agenda after the release of version 10.

Both the Calculation and Time tools have a basic workflow established. First, the semantic level of the STK Objects should be defined. Secondly, an ontological relationship is formed using Access, Deck Access, Chains, or a form of Coverage. After the STK Engine has computed the relationship, the Calculation and Time tools are used to organize and extend the information from the data provider components based on the definitions of the STK Objects and the ontological relationships.

### Calculation Tool

The Calculation Tool uses the data providers from the STK Objects and the STK Tool to control the algorithm components at a semantic level, giving users the ability to create more refined computations based on formed ontologies. In addition, using the Calculation tool, new components can be

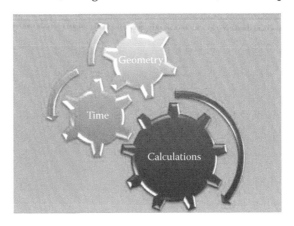

**FIGURE 19.3**
Geometry, Time, and Calculations.

added to the STK Objects and STK Tools to create more robust algorithms for an improved output. This capability allows endless possibilities for physics-based calculations. Furthermore, this also allows you to debug and validate the algorithms used for professional presentations and analysis.

## Time Tool with Timeline View

The Time tool leverages the Time data providers within the STK Objects and STK Tools without going into the Reports and Graphs area. If we look at an STK Object, the time interval is defined within the Basic Properties pages. The Time tool is primarily used after the STK Object is defined and a relationship using an STK Tool is established. For most of the ontological tools, the Access algorithm is used. The Access algorithm is time-based on valid time intervals of intervisibility. Any time-based component or subcomponent of an object or tool within the given relationship can be used within the Time tool. The data providers used with the Time tool gives you the ability to the control the data provider's component for the Access intervals established. These components are based on the semantic-level attributes and constraints defined within the STK Object and the STK Tool. The Time tool customizes how the data are computed and presented.

# Section IV

# Output

# 20

## Output

---

### Objectives of This Chapter

- Define Output
- Overview of Graphs and Reports
- Overview of Data Providers

---

### STK Output

STK output is the last part of the ontological study. The results of the relationship interactions are delivered in the form of reports, graphs, and animations. With each of these output forms, we also have access to the data providers from the given object and tool interaction computed. Here, like the rest of the book, the focus is not on "how to create output" but on how it affects the ontological study per se. Output is used to visualize results and to analyze the results for further refinement.

Let's look at a simple Access. We will use a simple exercise found in the Fundamental Training Guide at AGI entitled "Where Is the ISS?" If you go to the AGI.com website and search, you can retrieve this lesson online free of charge. As a high-level overview, let's take a look at the exercise. We can have one object, say a person with a telescope standing at Exton, Pennsylvania (or whatever city you choose). The second object would be the ISS. If we computed for Access and asked for a report, we would get a page of times that the ISS could be seen from the position of Exton. It would give us Start Time, Stop Time, and Duration of the intervisibility between the two objects. Or, if we ran an Azimuth-Elevation-Range (AER) report, we could even tell what the look angle (the azimuth and elevation angles) were and the distance with the time we should be looking. The simple calculations are from an algorithm that includes the start time and stop time of each object, the positions of each object, and the distances they were from each other. Since

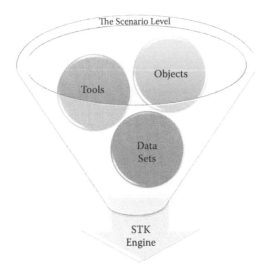

**FIGURE 20.1**
Ontology in STK.

Access is valid during points of intervisibility, then those would be the times we would consider.

The algorithm that is used to evaluate intervisibility is much like any other algorithm and is made of terms and operations. The terms are grouped by function and considered data providers within the software. If you go to the Custom Graphs and Charts area in version 9.*n* or use version 10.*n* software with the new Calculation and Time tools, you have easy access to the data providers. You can use these providers to create customized graphs, charts, and other forms of output.

## Graphs and Reports

Graphs and reports are generated after computing in either a static or dynamic strip-chart format. They have a standard quick-reporting retrieval from the Access button or you can create custom reports and graphs in the Reports and Graphs Manager. Both reports and graphs compute the ontological relationship developed within the tools. For instance, Access has a compute button for either graphs or reports. So does Chains. To compute Coverage, the Figure of Merit has a compute section after all the parameters are identified.

**The Ontological Relationship: Defined Objects and tools with output**

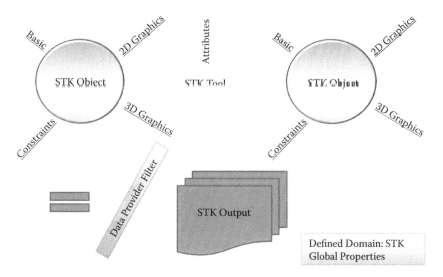

**FIGURE 20.2**
Full ontology.

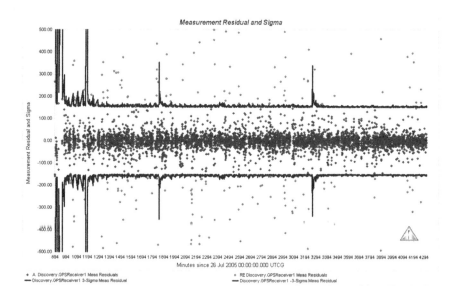

**FIGURE 20.3**
GPS residual.

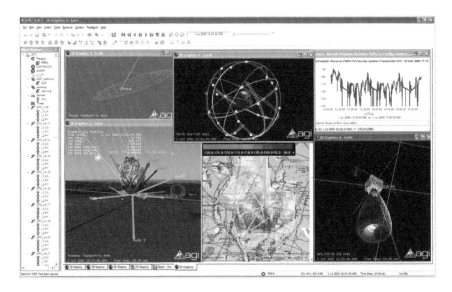

**FIGURE 20.4**
Graphs, Reports, and Dynamic Displays.

## Data Providers

STK Output is the last part of the ontological study. STK computes through the STK Engine. The results are in the form of graphs, reports, and animations for your analysis. Data providers are the segments, or terms, of the algorithms used to process the output. After calculations have processed, you have access to these data providers through tools. These data providers are the focus of the output for our ontological study in STK.

Because the output is already produced, the data providers act as a filter to show specified or isolated parts of the object's attributes and how they compute with the established relationship. Let's use a Communication object as an example.

We could use a satellite with a transmitter object and a facility with a sensor (as a pointing device) with a receiver. After developing the attributes for all the objects—Satellite (parent) and Transmitter (child) with a Facility (parent), Sensor (child), and Receiver (child's child)—then we could run a simple Access Link Budget. After running the link budget, we would be able to go into the Graphs and Reports Manager and review the data providers available to give us a more focused report instead of a full or detailed link budget. We could customize the graph to help us compare the signal-to-noise ratio to another link budget or even as the signal-to-noise ratio changes over time as the position of the satellite changes in relationship to the facility. Data providers allow us to filter out terms of the relationship and create a more focused study. When we get to this level of analysis, the use of developing your scenarios and refining your Spatial Temporal Information System with the process of an ontological study becomes even more apparent.

Data providers, as well as graphs, reports, and animations, are all part of the output in STK. They allow us to build a strong ontological study within STK that is usually reiterative in nature. The study is initialized by building your scenario from the concept, to the build-out of the objects, and then the build-out of the tools that define the relationships between the objects. After computing and then refining your analysis, refining the attributes of the objects and tools allows you to realize the concept of "video games for adults" in the form of mission modeling software. The use of ontology within the building of modeling software like STK is a unique approach. It requires a more formalized study of the use of analysis software such as a spatial temporal system. This book isn't designed to show you "how to"— that is what the AGI Comprehensive Training does, which is offered free at this time. However, it is here to help you understand, in a formal way, why the software behaves as it does. It allows you, as a user, to make the correct choice in your analysis endeavors.

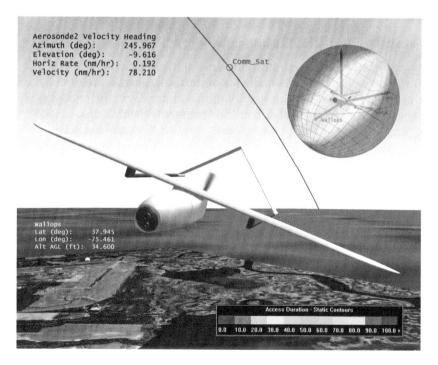

**FIGURE 20.5**
Wallops Island and Dynamic Display output.

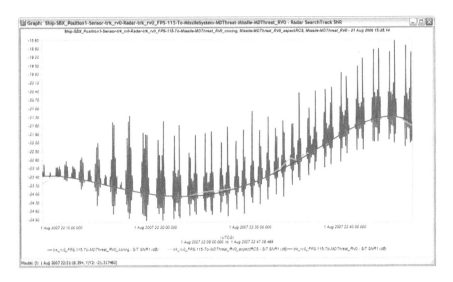

**FIGURE 20.6**
Radar and Missile Signal graph output.

## Closing Thoughts on Ontology

Ontology is an approach to studying spatial information systems. It allows you, the user, a method to realize a more formalized way to study your analysis. This works with most spatial information systems once you identify the objects and the mechanism used to create relationships. For STK, the mechanism is a visual tool, like Access, applied to create the relationships among objects. As the objects and tools are refined, the relationships become more meaningful.

# *References*

## Preface

Richardson, Douglas. 2011. *Space-Time Symposium,* opening plenary session, AAG, April 13.

## Chapter 1—Introduction

Aerospace Corporation. 2000. *Independent Verification and Validation (IV&V) for STK's High-Precision Orbit Propagation and Coordinate Frame Transformations.*

Analytical Graphics Inc. STK Comprehensive Course Work.

Analytical Graphics Inc. AGI STK Help files.

de Smith, Michael John, Michael F. Goodchild, and Paul A. Longley. 2006–2011. *Geospatial Analysis—A Comprehensive Guide.* 3rd ed. Matador on behalf of the Winchelsea Press.

Longley, Paul A., Michael F. Goodchild, David J. Maquire, and David W. Rhind. 2005. *Geographic Information Systems and Science.* 2nd ed. John Wiley & Sons.

McNeil, Linda McNeil. 2011. White paper on Spatial Temporal Analytics, presented to the AAG Spatial Temporal Symposium.

Roddick, John F., John F., Max Egenhofer, Erik Hoel, Dimitris Papadias, and Betty Salzberg. "Spatial, Temporal and Spatio-Temporal Databases—Hot Issues and Directions for PhD Research." Retrieved electronically from http://www.cs.ust.hk/~dimitris/PAPERS/SIGREC04.pdf.

## Chapter 2—Ontology

Analytical Graphics, Inc. 2012. System's Tool Kit Help file.

Analytical Graphics, Inc. 2012. STK Comprehensive Training Manual.

Arvidsson, F., and A. Flycht-Eriksson. "Ontologies I." 2008. http://www.ida.liu.se/~janma/SemWeb/Slides/ontologies1.pdf.

Celestrak.

DeNicola, Antonio, Michele Missikoff, and Roberto Navigli. 2009. "A Software Engineering Approach to Ontology Building." *Information Systems* (Elsevier) 34(2): 258–275. http://www.dsi.uniroma1.it/~navigli/pubs/De_Nicola_Missikoff _Navigli_2009.pdf.

Enderton, H. B. *A Mathematical Introduction to Logic*, 1st ed. San Diego, CA: Academic Press, 1972, p. 295. ISBN 978-0-12-238450-9; 2nd edition: 2001, ISBN 978-0-12-238452-3.

Gruber, Thomas R. 1993. "A Translation Approach to Portable Ontology Specifications." *Knowledge Acquisition*, 199–220.

Gruber, Thomas R. 1993. What Is Ontology? http://www-ksl.stanford.edu/kst/ what-is-an-ontology.html.

Gruber, Thomas R. 1995. "Toward Principles for the Design of Ontologies Used for Knowledge Sharing." *International Journal of Human-Computer Studies*, 907–928.

McNeil, Linda. 2011. *Spatial Temporal Analytics*. AG Spatial Temporal symposium.

Mizoguchi, R. 2004. "Tutorial on Ontological Engineering, Part 3: Advanced Course of Ontological Engineering." *New Generation Computing* (Ohmsha & Springer-Verlag) 22(2): 198–220.

Vallado, David A., and T. S. Kelso. "Using EOP and Space Weather Data for Satellite Operation."

## Chapter 3—The Scenario

Analytical Graphics, Inc. STK Comprehensive Manual.
Analytical Graphics, Inc. STK Help files.
McNeil, Linda. 2011. *Spatial Temporal Analytics*. AG Spatial Temporal symposium.

## Chapter 4—Objects

Analytical Graphics, Inc. ADF Administrator's Guide.
Analytical Graphics, Inc. STK Comprehensive Manual.
Analytical Graphics, Inc. STK Help Files.
Analytical Graphics, Inc. Working with an AGI Data Federate.
National Oceans and Atmospheric Administration. http://www.ngs.noaa.gov/.

## Chapter 5—Area Targets

Analytical Graphics, Inc. STK Help Files.
National Geospatial-Intelligence Agency (NGA).

## Chapter 6—Targets and Facilities

Analytical Graphics, Inc. STK Help Files.

## Chapter 7—Moving Objects

Linda McNeil. 2011. *Spatial Temporal Analytics.* AG Spatial Temporal symposium.

## Chapter 8—Aircraft

Analytical Graphics, Inc. AGI.com.
Analytical Graphics, Inc. STK Help Files.
http://www.agi.com/products/by-product-type/applications/stk/add-on-modules/stk-solis/default.aspx.
http://www.go-asi.com/software/stk-solis.html.
McNeil, Linda. 2011. *Spatial Temporal Analytics.* AG Spatial Temporal symposium.
*Summary Report, STK B-747 Aircraft Type Model Development Validation and Verification Contract Final Report #2.* 2005. Saab Sensis Corporation.

## Chapter 9—Satellites

Analytical Graphics, Inc. STK Help Files.
Kelso, T. S. http: Celestrak.com.

## Chapter 10—Advanced Satellites

Analytical Graphics, Inc. Advanced STK Astrogator Training Manual, Applied Defense Solutions.
Analytical Graphics, Inc. STK Comprehensive Manual.
Analytical Graphics, Inc. STK Help Files.
Berry, Matt. 2006. "Using STK/Astrogator as a Graphical Programming Language for Planning." AGI User Exchange.

Carrico, John, and Emmet Fletcher. "Software Architecture and Use of Satellite Tool Kit's Astrogator Module for Libration Point Orbit Missions." Analytical Graphics. http://astrogatorsguild.com/wp-content/papers/SoftwareArchitectureAnd Use_Astrogator.pdf.

## Chapter 11—Child Objects

Analytical Graphics, Inc. AGI.com
Analytical Graphics, Inc. Communications Course.
Analytical Graphics, Inc. STK Help Files.

## Chapter 12—Constellations

Analytical Graphics, Inc. STK Help Files
McNeil, Linda. 2011. *Spatial Temporal Analytics.* AG Spatial Temporal symposium.

## Chapter 13—STK Tools

Analytical Graphics, Inc. STK Help Files
McNeil, Linda. 2011. *Spatial Temporal Analytics.* AG Spatial Temporal symposium.

## Chapter 14—Access and Deck Access

Analytical Graphics, Inc. Light Time Delay and Apparent Position. (PDF found in Help page from Access Help)
Analytical Graphics, Inc. STK Help Files
Coppola, Vincent. "Interplanetary Computations: Light Time Delay and Line of Sight." AGI User Exchange. 2006.
Kelso, T. S. Celestrak.com.
http://www.russianspaceweb.com/phobos_grunt.html.
http://celestrak.com/events/reentry/phobos-grunt.asp.
http://www.agi.com/agiforum/messages.aspx?topicid=772.
http://www.agi.com/downloads/events/2011-singapore-summit/STK-10-Overview.pdf.
http://www.astro.ucla.edu/~wright/deflection-delay.html.

The Time tool (version 10) http://www.youtube.com/watch?v=JtSG8h-a-UQ&feature =related.

Thorstensen, John. "Coordinates, Time and the Sky." Department of Physics and Astronomy, Dartmouth College, Hanover, NH.

Young, Andrew T. http://mintaka.sdsu.edu/GF/explain/atmos_refr/dip.html.

## Chapter 15—Chains

Analytical Graphics, Inc. STK Help Files

Analytical Graphics, Inc. Light Time Delay and Apparent Position. (PDF found in help page from Access Help).

Analytical Graphics, Inc. STK Web Online Help System. http://www.agi.com/resources/help/online/stk/source/stk/importfiles-04.htm.

http://www.agi.com/downloads/events/2011-singapore-summit/STK-10-Overview.pdf

McNeil, Linda. 2011. *Spatial Temporal Analytics.* Analytical Graphics. AAG Symposium.

## Chapter 16—Coverage

Analytical Graphics, Inc. STK Help Files.

McNeil, Linda. 2011. *Spatial Temporal Analytics.* Analytical Graphics. AAG Symposium.

## Chapter 17—Communications

Analytical Graphics, Inc. Communications Course.

Analytical Graphics, Inc. Light Time Delay and Apparent Position. (PDF found in Help page from Access Help).

Analytical Graphics, Inc. STK Help Files.

http://cp.literature.agilent.com/litweb/pdf/5965-7160E.pdf.

http://docwiki.cisco.com/wiki/Internetworking_Technology_Handbook.

http://technav.ieee.org/tag/2758/digital-communication.

http://www.agi.com/downloads/events/2011-singapore summit/STK-10-Overview.pdf.

http://www.alionscience.com.

http://www.complextoreal.com/tutorial.htm.

http://www.fab-corp.com/pages.php?pageid=2.

http://www.intelsat.com/resources/tech-talk/eclipse-seasons.asp.

McNeil, Linda. 2011. *Spatial Temporal Analytics.* Analytical Graphics. AAG Symposium.

## Chapter 18—CAT

Analytical Graphics, Inc. STK Help Files.
Kelso, T. S. http: Celestrak.com

## Chapter 19—Nonontological Tools

Tanygin, Sergei. Calculation Tool webinar.
Tanygin, Sergei. Time Tool and Timeline View webinar.
Tanygin, Sergei. Vector Geometry Tool. AGI User Exchange 2006.

## Appendix A—Plug-In Scripts

Permission granted by Analytical Graphics, Inc., www.agi.com

## Appendix B—Light Time Delay and Apparent Position

Permission granted by Analytical Graphics, Inc., www.agi.com
Seidelmann, Ken, ed. 1992. *Explanatory Supplement to the Astronomical Almanac.*

## Appendix C—Transmitter GUI Flow Chart

Permission granted by Analytical Graphics, Inc. www.agi.com. Permission Granted
    by Analytical Graphics, Incorporated

# Appendix A: Plug-in Scripts

Analytic Graphics Inc.

## Introduction

Plug-in Scripts provide a method for incorporating customer-specific, non-generic modeling into STK computations. Users create scripts (written in either MATLAB, Perl, or VBScript) that are executed at specified times during a computation to integrate their own models within STK's computational framework. Plug-in Scripts provide a simple-to-use mechanism for customizing STK while leveraging its generic computational algorithms.

NOTE: Python version 2.4 is also supported.

For example, consider modeling the forces on the International Space Station (ISS). The ISS has large solar panel arrays that rotate to align with the sun as it travels in its orbit. Thus, the area of the vehicle that is exposed to atmospheric drag varies periodically over each orbit due to the changing orientation of the solar panels. Normally, STK/Astrogator assumes a constant area when computing the drag force (like almost all orbit propagators). It is not that the computations involved with varying drag area are difficult—they are not—rather, the reason that most integrators don't handle varying area is that there is no generic model of variable satellite area for a user to choose. Every satellite's drag area is very much dependent on its moving surfaces, their attitudes, and the satellite's own attitude: that is, it is highly specific to the satellite and not generic at all.

How can you customize your models without rewriting the software for each case?

Answer: By using Plug-in Scripts. The user can create a Plug-in Script that just models the surface area as a function in time and have STK/Astrogator calculate the drag force using this new surface area model. Because the user has complete control over the script, the model can be as complicated or as simple as the user needs it to be. The remaining aspects of integrating the orbit ephemeris (computing the drag force based on the area computed within the script, computing other forces, numerically integrating the equations of motion, maintaining accuracy using a step error control method, etc.) are handled automatically by STK. This allows the user complete flexibility in modeling just the task at hand without requiring the development and testing of other supporting code.

Each Plug-in Script chooses its own inputs and outputs from a group of available variables. The outputs of a Plug-in Script define its intended use: for example, a script that outputs force model parameters can be used as a Propagator Plug-in Script but not as an Access Constraint Plug-in Script. Because each Plug-in Script has a very specific task, its possible outputs have been limited by design (but even then there are still quite a few). The inputs to a script are another matter entirely: calling user created scripts during computations would be of little use if the inputs that were needed to create the outputs were unavailable. The Plug-in Points described below provide a very large number of possible inputs: in many cases, virtually all ephemeris and attitude-related data from any object in the scenario can be accessed as an input.

By providing all the required input data and leveraging existing computational algorithms, Plug-in Scripts provide a simple, elegant, quick solution for customizing models within STK.

The ISS example above is specific to one particular problem. Plug-in Scripts can be used in many other computations to provide customization of many different models. Plug-in Scripts are available for each of the areas shown in Table 1.

TABLE 1

Plug-in Script Applications

| Application | Tasks Performed by Scripts |
|---|---|
| Astrogator Propagator | Modeling additional forces (e.g., lift), atmospheric density, time-varying parameters. |
| Astrogator Engine Model | Modeling engine thrust, mass flow rate, and Isp. |
| Astrogator Calc Objects | Modeling a user-specific computation. |
| Vector Geometry Tool Custom Vector | Modeling a time-varying vector. |
| Vector Geometry Tool Custom Axes | Modeling time-varying axes. |
| Access Constraint | Modeling a visibility constraint that is utilized during access computations. |
| Attitude Simulator | Modeling external torques and implementing attitude control laws. |

## Installation

The following sections contain installation instructions for MATLAB, VBScript, and Perl users.

### Installation for MATLAB Users

STK Plug-in Scripts require MATLAB 6. To use MATLAB scripts, one must first perform the STK/MATLAB installation. After installation, run MATLAB and, at the command prompt, run agiInit. Then follow the steps to complete the installation.

> NOTE TO PC USERS: It is important that the MATLAB executable and dll's be on the user's path environmental variable. The installation process for MATLAB 6 does this; however, if the user has uninstalled a previous version of MATLAB, the old MATLAB path information may still be present in the path environmental variable. Moreover, the old dll's may still be present on the user's system even after uninstall. If the old path information and the old dll's are present on your system, then the proper MATLAB 6 versions may not be accessible from STK. To remedy this problem, you will need to manually change the path environmental variable to point to the MATLAB 6 installation.

The MATLAB debugger can be used to debug a MATLAB function when the function is called directly from MATLAB itself; however, it is not compatible with functions called from outside MATLAB: using the MATLAB debugger with STK MATLAB Plug-in Scripts can cause both programs to hang or to crash.

NOTE: If MATLAB was started by STK, then the proper manner to close MATLAB is also from STK. To close MATLAB, open the application Preferences from the Edit: Preferences menu item, go to the MATLAB page, and click the Disconnect button to close MATLAB. In fact, the quit command in MATLAB will not quit MATLAB if STK has opened it. The only way to quit MATLAB once started by STK is to quit STK, use the Disconnect button as described above or to use the command "quitforce" at the MATLAB prompt.

## Installation for VBScript Users

VBScript is automatically installed with Microsoft Windows and is available free. There is no further installation step required.

The VBScript functionality in Windows is enabled through the file atl.dll. To use STK Plug-in Scripts, you must have a copy of VBScript delivered with Windows NT (or later). Users who have upgraded from Windows 98 to Windows NT, 2000, or XP may not have upgraded to the correct version of atl.dll. The latest version can be downloaded free of charge from Microsoft at its website:

```
http://activex.microsoft.com/controls/vc/atl.cab
```

Once the new software is installed, you need to register AgScript.dll as a COM server. Open a Command Prompt and change directories until you are in the directory containing the STK executables and dll's. The AgScript.dll is in this directory. Now run:

```
regsvr32 AgScript.dll
```

One must be alert to distinguish between VBScript (Visual Basic Scripting Edition), VB (Visual Basic), and VBA (Visual Basic for Applications). Some keywords and functionality available in VB/VBA are not available in VBScript. Moreover, several VBScript books document features that are not available in VBScript (but are available in VB/VBA). Users may want to purchase a VBScript book that is careful about this distinction.

NOTE: If you use the MS Script Debugger, which Microsoft makes available free of charge to debug VBScript scripts (see http:/msdn/microsoft. com/scripting), a syntax error in the script will cause the debugger to stop at that error. However, in general, the MS Script Debugger cannot be used to debug scripts while they are being used by STK.

## Installation for Perl Users

STK Plug-in Scripts require version 5.6.1 of Perl, which is available free. For PC users, the version of Perl from ActiveState (www.activestate.com) is required. Perl Plug-in Scripts were developed using build 628 of ActiveState's version of Perl—later builds should be compatible. Contact AGI at support@ stk.com if you need assistance.

Please note that you may need to set an environmental variable to enable STK Plug-in Scripts to work with Perl. If you experience errors trying to use Perl (with a message indicating that a file could not be found anywhere along the include path variable given by @INC), set a system environmental variable AGPERL_LIB_PATH to the following:

```
.;[Perl_Install_Directory]/lib;[Perl_Install_Directory]/site/
lib
```

where [Perl_Install_Directory] is the directory where Perl is installed on your computer. Directories (separated by semicolons on a PC, separated by colons on UNIX) in the variable AGPERL_LIB_PATH are added to the include path @INC when Perl is started. On UNIX, AGPERL_LIB_PATH should be set in the user's.cshrc file.

After installing the correct version of Perl, you will need to copy the Perl module STKUtil.pm from the <STK Home>/Connect/PERL_LIB directory into a directory that is on the Perl library search path (i.e., a directory path contained in @INC). Typically, users will copy the file to [Perl_Install_Directory]/ lib, where [Perl_Install_Directory] is the directory where Perl is installed on your computer. The STKUtil module exports several utility functions that users may want to use in their own scripts. To use the module, include the line

```
use STKUtil qw (printOut getInputArray);
```

near the top of the file. The getInputArray function is a user convenience function for translating between references to an array and the array itself: this is useful for dealing with inputs to Perl scripts from STK (see the section of Function Signatures for Perl). The printout function can be used to print out lines of text; on a PC, it pops up a message box containing the text that the user must dismiss.

> NOTE: On PC only, you cannot use standard input (stdin), standard output (stdout), and standard error (stderr) to print out any messages. Instead you can print to a file or use printout [or use Win32::MsgBox() directly] to pop messages up in a dialog box on the screen.

## Scripting Environments

By convention, the extension of the filename determines the language to be used (see Table 2).

**TABLE 2**

Languages and Extensions

| Extension | Language |
| --- | --- |
| .m | MATLAB |
| .dll | MATLAB compiled code |
| .vbs | VBScript |
| .pl | Perl |

The name of the function that is called by a Plug-in Script is based on the file-name. For example, the name of the function called when using the filename

```
C:\stk\user1\VB_ForceModel_Eval.vbs is VB_ForceModel_Eval.
```

> NOTE: This convention is the same as the MATLAB convention for dealing with user-defined junctions (i.e., .mfiles). This convention was chosen to avoid name collisions since MATLAB and VBScript do not understand the concept of namespaces.

## MATLAB Environment

Only one copy of MATLAB is opened by STK for use with Plug-in Scripts. All MATLAB Plug-in Scripts share the same workspace environment. STK does not connect to MATLAB until some MATLAB Plug-in Script attempts to execute a script. This may occur during application start-up or when loading scenario files, depending on the Plug-in Point being used.

## VBScript Environment

Only one copy of a VBScript engine is opened by STK for use with VBScript Plug-in Scripts. All VBScript Plug-in Scripts share the same workspace environment. STK does not create a VBScript environment until some VBScript Plug-in Script attempts to execute a script. This may occur during application start-up or when loading scenario files, depending on the Plug-in Point being used.

As part of the creation process, all VBScript files (i.e., files with extension .vbs) within certain directories are automatically loaded into the workspace. Users may add their own VBScript files to these directories and they will be automatically loaded as well. The directories are:

(i) <STK install folder>/STKData/Scripting/Init

(ii) <STK user area>/Scripting/Init

where <STK install folder> refers to the directory path that is the parent of STKData, and <STK user area> refers to the user's configuration directory. A success or failure message will be written for each of the files found and loaded.

> NOTE: This mechanism allows users to create certain VBScript utilities that other VBScript scripts may utilize, since there is no internal mechanism in VBScript to load and execute another script.

## Perl Environment

Only one copy of the Perl interpreter is opened by STK for use with Perl Plug-in Scripts. All Perl Plug-in Scripts share the same workspace environment. STK does not create a Perl environment until some Perl Plug-in Script attempts to execute a script. This may occur during application start-up or when loading scenario files, depending on the Plug-in Point being used.

As part of the creation process, all Perl files (i.e., files with extension.pl) within certain directories are automatically loaded into the workspace. Users may add their own Perl files to these directories, and they will be automatically loaded as well. The directories are:

(i)   <STK install folder>/STKData/Scripting/Init
(ii)  <STK user area>/Scripting/Init

where <STK install folder> refers to the directory path that is the parent of STKData, and <STK user area> refers to the user's configuration directory. A success or failure message will be written for each of the files found and loaded.

## Function Signatures

Every script function has the same signature, takes one input argument, and returns one output argument. In VBScript and Perl, the input and output arguments are arrays; in MATLAB, each argument is a MATLAB Struct.

Each script function is called for at least two purposes:

1. to register a list of requested inputs and outputs for the *compute* function; and
2. to compute the outputs based upon the inputs.

The registration process for a Plug-in Script occurs as an initialization step; if the script successfully initializes, then it can provide the requested inputs and handle the requested outputs when asked to compute. The list of requested inputs and outputs cannot be changed until another initialization step is performed. To aid computational efficiency, the initialization process for a Plug-in Script is performed as infrequently as possible: some Plug-in Scripts initialize only when the STK application starts up (e.g., some calls made for Access Constraint Plug-in Points); some reinitialize only when the user requests them to be reinitialized (e.g., Vector Geometry Tool Plug-in Points); others reinitialize every time a computational process starts (Astrogator Plug-in Points).

The available input and output arguments depend on the Entry Point of the Plug-in point itself. Some Entry Points have no inputs or outputs, but others can have many inputs and outputs. See the documentation for each Entry Point of each Plug-in point for the list of available inputs and outputs.

> **NOTE:** The units for inputs and outputs are STK internal units. In most cases, these units are standard SI units such as meter, kilogram, second, etc. An exception is Power, for which the unit is dbW (decibel Watts). Note that angles are in radians.

Examples of each language's function signature are given below. Note that a condition is checked to determine whether the script function is to return a list of descriptors or is to compute the outputs on the basis of the input values.

## MATLAB Functions

### *Function Signature*

As an example, consider the file MATLAB_CalcObject_Eval.m, which contains the following:

```
function [output] = Matlab_CalcObject_Eval (input)
switch input.method
  case 'register'
    % create output here to register inputs/outputs
  case 'compute'
    computeData = input.methodData;
      % create output struct containing values for the outputs
        based on the
      % inputs in computeData
    otherwise
      output = [];
end
```

Every MATLAB.m file must follow this same paradigm. Note that input is a MATLAB Struct whose method value is a string (the calling mode). On compute, the inputs are to be extracted from computeData for use in computing output.

### Registration of Input and Output Arguments

Each script function requests inputs and outputs by returning a descriptor for each of the requested arguments. The descriptor contains a list of strings of the form "keyword, value". STK parses the keyword-value pairs to identify the requested argument. The number of keyword-value pairs that is required to describe an argument depends upon the complexity of the argument: simple inputs and outputs require fewer pairs than more complicated ones. If a requested argument cannot be identified, then that script function is disabled.

> **NOTE:** The keyword string is case insensitive—we capitalize certain letters for legibility in examples and documentation. The value string is case sensitive, however, with some exceptions (e.g., ArgumentType values of "Input" and "Output" are actually case insensitive).

Every requested input or output argument descriptor must contain the keywords ArgumentType and ArgumentName. The value of ArgumentType is either Input or Output. The value of ArgumentName can be any user-specified variable name that obeys the language's syntax for a valid variable name. (Thus, special characters and spaces are not allowed.) The ArgumentName is intended to be a unique name for that argument when referenced in the script.

> **NOTE:** The order of the keyword-value pairs within an argument descriptor is arbitrary, although we adopt a convention where ArgumentType is first, so that you can identify which arguments are inputs and outputs more readily.

Argument descriptors in MATLAB are specified in the form of a MATLAB CellArray. For example, the code:

```
incDescriptor = {'ArgumentType', 'Input',...
                 'ArgumentName', 'inc',...
                 'Name', 'Inclination',...
                 'Type', 'CalcObject'};
```

describes an Input argument which is a CalcObject named Inclination, which will be referred to in this .m file by the name 'inc'. When the script is called in compute mode, the name 'inc' will be used to extract the input value for this argument.

Here is an example showing three arguments (one output, two inputs) being registered:

```
case 'register'
  valueDescriptor = {'ArgumentType', 'Output',...
                     'ArgumentName', 'value',...
                     'Name','Value'};
  incDescriptor = {'ArgumentType', 'Input',...
                   'ArgumentName', 'inc',...
```

```
                    'Name', 'Inclination',...
                    'Type', 'CalcObject'};
  rightAscDescriptor = {'ArgumentType', 'Input',...
                       'ArgumentName', 'rightAsc',...
                       'Name', 'RAAN,'...
                       'Type', 'CalcObject'};
  output = {valueDescriptor, incDescriptor,
            rightAscDescriptor};
```

### Computing Outputs from Inputs

When the script is called upon to compute, the inputs and outputs are handled as follows:

```
case 'compute'
computeData = input.methodData;
output.value = cos(computeData.inc)*cos(computeData.rightAsc);
```

Note that the ArgumentNames are used to extract and set the data.

> NOTE: Arrays in MATLAB are passed as column vectors, not row vectors.

## VBScript Functions

### Function Signature

Script functions using VBScript are always called with one input array. If the first argument in the array is not empty, then it is a String containing the name of the calling mode. The calling mode is set to register when the script function is supposed to return an array of requested inputs and outputs; it is usually set to vbEmpty to indicate to the script that it is to compute outputs based on any inputs given in the array, though the calling mode may be set to compute. As an example consider the file VB_CalcObject_Eval.vbs, containing the following:

```
Function VB_CalcObject_Eval (argArray)
Dim retVal
If IsEmpty(argArray(0)) Then
    'do compute
    retVal = VB_CalcObject_Eval_compute(argArray)
ElseIf argArray(0) = "register" Then
    retVal = VB_CalcObject_Eval_register()
ElseIf argArray(0) = "compute" Then
    'do compute
    retVal = VB_CalcObject_Eval_compute(argArray)
Else
    'bad call
```

```
    retVal = Empty
    End If
    VB_CalcObject_Eval = retVal
End Function
```

Note that the registration and compute aspects of the VB_CalcObject_Eval function are handled by other VBScript functions: VB_CalcObject_Eval_register and VB_CalcObject_Eval_compute. You can define these functions in the same file; this programming style leads to script readability.

> NOTE: VBScript does not understand the concept of namespaces. This means that if two separate VBScript files that define exactly the same function name are used sometime during the STK session, then the first version of the function is overridden by the second version. Most probably, that type of behavior is not the desired one. To avoid this problem, we *highly recommend* that all functions and any global variables use the filename as a prefix. That is why the register function for VB_CalcObject_Eval (above) is not named simply "register" (which could be overridden by some other file that defines a register function) but instead "VB_CalcObject_Eval_register" (which is much less likely to be overridden in a different file that obeys this convention).

When the script performs its *compute* function, the values contained in the input array are in the same order as that in which the Inputs were registered, with the first Input at index 1, the second at index 2, etc., and with the calling mode occupying index 0 of the array.

### Registration of Input and Output Arguments

Each script function requests inputs and outputs by returning a descriptor for each of the requested arguments. The descriptor contains a list of strings of the form "keyword = value". STK parses the keyword-value pairs to identify the requested argument. The number of keyword-value pairs that is required to describe an argument depends upon the complexity of the argument: simple inputs and outputs require fewer pairs than more complicated ones. If a requested argument cannot be identified, then that script function is disabled.

> NOTE: The keyword string is case insensitive—we capitalize certain letters for legibility in examples and documentation. The value string is case sensitive however, with some exceptions (e.g., ArgumentType values of "Input" and "Output" are actually case insensitive).

Every requested input or output argument descriptor must contain the keywords ArgumentType and ArgumentName. The value of ArgumentType is either Input or Output. The value of ArgumentName can be any user-specified variable name that obeys the language's syntax for a valid variable name. (Thus, special characters and spaces are not allowed.) The ArgumentName is intended to be a unique name for that argument when referenced in the script.

> **NOTE:** The order of the keyword-value pairs within an argu-
> ment descriptor is arbitrary, although we adopt a convention where
> ArgumentType is first, so that you can identify which arguments are
> inputs and outputs more readily.

During the registration process, an array of descriptors must be generated
and returned. Each descriptor can be either an array of Strings, where each
String is a keyword-value pair, or a String containing keyword-value pairs
separated by semicolons. An example follows:

```
Function VB_ForceModel_Eval_register ()
   ReDim descripStr(3), argStr(4)
   descripStr(0) = "ArgumentType = Output"
   descripStr(1) = "Name = Status"
   descripStr(2) = "ArgumentName = Status"
   argStr(0) = descripStr
   Dim singleLineDescripStr
   singleLineDescripStr = "ArgumentType = Output; Name =
      Acceleration;"
   singleLineDescripStr = singleLineDescripStr & _
   "RefName = LVLH; ArgumentName = accel"
   argStr(1) = singleLineDescripStr
   ReDim descripStr(4)
   descripStr(0) = "ArgumentType = Input"
   descripStr(1) = "Name = Velocity"
   descripStr(2) = "RefName = Inertial"
   descripStr(3) = "ArgumentName = Vel"
   argStr(2) = descripStr
   argStr(3) = "ArgumentType = Input; Name = DateUTC;
      ArgumentName = Date"
   VB_ForceModel_Eval_register = argStr
End Function
```

## Computing Outputs from Inputs

The order of registration determines the order of the arguments in the
incoming and outgoing arrays of the requested data. For example, based on
the registration given above, there are two requested inputs. When called
upon to compute, the input array will then consist of three elements:

Index 0: The calling mode, usually *vbEmpty*

Index 1: The velocity (an array of three doubles), referred to by the
name *Vel* in the script

Index 2: The UTC date (a String), referred to as *Date* in the script.

Accordingly, the output array consists of two elements:

Index 0: Status (a String)

Index 1: The acceleration (an array of three doubles) relative to the LVLH frame, referred to as *accel* in the script.

An error results if:

- an array is not returned,
- fewer outputs were returned than were requested,
- the incorrect type of data is returned for an argument (e.g., a String was returned where a Double was expected), or
- an array is returned that is too short (e.g., returning two doubles rather than the expected three).

If an error results, the script is turned off, and a message explaining the problem is output.

> **NOTE:** If the script function returns a String instead of an array, an error results which then will cause the script to be turned off. Additionally, the returned string will be written to the message window. Thus, the script has the ability to turn the script off and indicate a reason for doing so, simply by returning a String.

### Accessing Data on the Basis of ArgumentName

Because the order of registration implies the order of the data in the input array and the output array, it is imperative that the correct association between the argument and its index number be used. Thus, adding new arguments in the middle of the current list or even just reordering the list of descriptors in the register function will require modifications of the index numbers used in the *compute* function. To alleviate this burden somewhat, the user can choose to use auxiliary classes that hold the association between argument and index number. This allows the user to access data by name, which should result in fewer modifications being necessary in the *compute* function when the registration function is altered.

After the registration process occurs, STK makes an additional call to VBScript to create two Classes (one for inputs, one for outputs) that can be used to access the data by ArgumentName (rather than using index numbers). These classes can be created and stored during the first call to compute, and then accessed by compute thereafter. The classes are created by calling the function g_GetPlug-inArrayInterface. For example, consider the function VB_ForceModel_Eval. Its *compute* function would look like this:

```
Function VB_ForceModel_Eval_compute(stateData)
  'This function ASSUMES that VB_ForceModel_Eval_register
  will set
```

```
   'VB_ForceModel_Eval_globalVar to -1
 If VB_ForceModel_Eval_globalVar < 0 Then
    'If one were to uncomment certain statements below, then a
    description of
    'the Inputs and the Outputs for this function would be
popped up in a
    'MsgBox and the user would need to hit 'OK'
    'Dim outStr
    Set VB_ForceModel_Eval_Inputs = _
      g_GetPlug-inArrayInterface("VB_ForceModel_Eval_Inputs")
    'outStr = VB_ForceModel_Eval_Inputs.Describe ()
    'MsgBox outStr
    Set VB_ForceModel_Eval_Outputs = _
      g_GetPlug-inArrayInterface("VB_ForceModel_Eval_Outputs")
    'outStr = VB_ForceModel_Eval_Outputs.Describe ()
    'MsgBox outStr
    'MAKE sure this If-Then block is executed only once
      '(unless VB_ForceModel_Eval_globalVar was reset to -1 by
      'VB_ForceModel_Eval_register
      VB_ForceModel_Eval_globalVar = 1
 End If
 'Declare some temporary variables
 Dim factor, cbiVel, cbiSpeed, dateString
 dateString = stateData(VB_ForceModel_Eval_Inputs.Date)
 factor = 0.000001
 cbiVel = stateData(VB_ForceModel_Eval_Inputs.Vel)
 cbiSpeed =
sqr (cbiVel (0) *cbiVel (0) + cbiVel (1) *cbiVel (1) +cbiVel
(2) *cbiVel (2))
 Redim accelValue (3)
 accelValue (0) = 0.0
 accelValue (1) = factor*cbiSpeed
 accelValue (2) = 0.0
 'Declare an array of the proper size
 Dim returnValue (2)
 'Assign output values
 returnValue (VB_ForceModel_Eval_Outputs.accel) = accelVal
 returnValue (VB_ForceModel_Eval_Outputs.Status) = "Still Okay"
 VB_ForceModel_Eval_compute = returnValue
End Function
```

The variables VB_ForceModel_Eval_globalVar, VB_ForceModel_Eval_ Inputs, and VB_ForceModel_Eval_Outputs are global variables and must be declared by a Dim statement outside of any function. During VB_ ForceModel_Eval_register, VB_ForceModel_Eval_globalVar should be set to –1. The first time VB_ForceModel_Eval_compute is called, the If-Then statement is executed, causing two classes to be created (VB_ForceModel_ Eval_Inputs and VB_ForceModel_Eval_Outputs). Because they have been declared global, they are remembered for the next call to compute.

Note that the inputs are accessed from stateData using the class VB_ForceModel_Eval_Inputs and the ArgumentName that was registered. Similarly, the outputs are assigned into the returned array using the class VB_ForceModel_Eval_Outputs and the ArgumentName that was registered. The classes provide a mechanism in the script to access the inputs and outputs by name, rather by index number. Of course, they do this by mapping the ArgumentName to an index value:

```
Name                                        Index Value
VB_ForceModel_Eval_Inputs.Vel                    1
VB_ForceModel_Eval_Inputs.Date                   2
VB_ForceModel_Eval_Outputs.Status               0
VB_ForceModel_Eval_Outputs.accel                 1
```

(Remember, on input the argument at index 0 is the calling mode.)

> **NOTE:** A matrix is passed in VBScript as a single array, containing the rows of the matrix in order, not in the form of an array of arrays.

> **CAUTION:** The user must be careful to respond to all message boxes initiated by VBScript. Suppose a computation is started that uses a VBScript function that pops up a message box. If, instead of responding to the message, the user clicks the Cancel button on the progress bar to stop the computation, the message box may be left stranded: dismissing it or calling a VBScript function again can cause STK to crash. Before using the cancel button, be sure to dismiss all popped up messages.

## Perl Functions

### *Function Signature*

Script functions using Perl are always called with one input array. If the first argument in that array is not undefined, then it is a String containing the name of the calling mode. The calling mode is set to register when the script function is supposed to return an array of requested inputs and outputs; it is usually set to undefined to indicate to the script that it is to compute outputs based on any inputs given in the array, though the calling mode may be set to compute. As an example consider file Perl_CalcObject_Eval.pl, containing the following:

```
sub Perl CalcObject Eval
{
  # the inputs to the script arise as a reference to an array
  # the STKUtil::getInputArray function is used to get at the
  array itself
  my @stateData = STKUtil::getInputArray(@_);
  my @retVal;
  if (!defined($stateData[0]))
  {
```

```
    # do compute
    @retVal = Perl_CalcObject_Eval_compute(@stateData);
    }
    elsif ($stateData[0] eq 'register')
    {
      @retVal = Perl_CalcObject_Eval_register();
        }
      elsif ($stateData[0] eq 'compute')
      {
        @retVal = Perl_CalcObject_Eval_compute(@stateData);
        }
      else
      {
        # error: do nothing
      }
  # MUST return a reference to an array, as shown below return
  \@retVal;
}
```

> **NOTE:** To check for compilation errors, we have adopted the Perl convention that all Perl scripts should end with "1;"

Note that the registration and compute aspects of the Perl_CalcObject_Eval function are handled by other Perl functions: Perl_CalcObject_Eval_register and Perl_CalcObject_Eval_compute. You can define these functions in the same file—this programming style leads to script readability.

> **NOTE:** Although Perl does understand the concept of namespaces, many Perl programmers are unaware of them, and so we have not assumed such knowledge in developing the call signature above. As an alternative to using namespaces, functions and global variables are named in a manner to prevent name collision in different Perl files. For example, if two separate Perl files that define exactly the same function name are used sometime during the STK session, then the first version of the function is overridden by the second version. Most probably, that type of behavior is not the desired one. To avoid this problem, we *highly recommend* that all functions and any global variables use the filename as a prefix: that is why the register function for Perl_CalcObject_Eval (above) is not named simply "register" (which could be overridden by some other file that defines a register function) but instead Perl_CalcObject_Eval_register (which is much less likely to be overridden in a different file that obeys this convention).

When the script performs its *compute* function, the values contained in the input array @stateData are in the same order as that in which the Inputs were registered, with the first Input at index 1, the second at index 2, etc., and with the calling mode occupying index 0 of the array.

> **NOTE:** Perl handles a function argument that is to be "one input argument that itself is an array" by passing a reference to an array. Thus, STK

will call the Perl function by passing to it a reference to an array, and expect from the script a reference to an array. Since references may not be familiar to all users, we have created a helper function, STKUtil::getInputArray, which dereferences the incoming reference into the array itself: see how @stateData is created. Once the array is known, Perl is used in a more normal manner wherein arrays themselves are passed as arguments and outputs. On output back to STK, however, one must return a reference to the array (\@retVal) and not the array itself (@retVal).

### Registration of Input and Output Arguments

Each script function requests inputs and outputs by returning a descriptor for each of the requested arguments. The descriptor contains a list of strings of the form "keyword = value". STK parses the keyword-value pairs to identify the requested argument. The number of keyword-value pairs that is required to describe an argument depends upon the complexity of the argument: simple inputs and outputs require fewer pairs than more complicated ones. If a requested argument cannot be identified, then that script function is disabled.

> NOTE: The keyword string is case insensitive—we capitalize certain letters for legibility in examples and documentation. The value string is case sensitive, however, with some exceptions (e.g., ArgumentType values of "Input" and "Output" are actually case insensitive).

Every requested input or output argument descriptor must contain the keywords ArgumentType and ArgumentName. The value of ArgumentType is either Input or Output. The value of ArgumentName can be any user-specified variable name that obeys the language's syntax for a valid variable name. (Thus, special characters and spaces are not allowed.) The ArgumentName is intended to be a unique name for that argument when referenced in the script.

> NOTE: The order of the keyword-value pairs within an argument descriptor is arbitrary, although we adopt a convention where ArgumentType is first, so that you can identify which arguments are inputs and outputs more readily.

During the registration process, an array of descriptors must be generated and returned. Each descriptor can be either an array of Strings, where each String is a keyword-value pair, or a String containing keyword-value pairs separated by semicolons. An example follows:

```
sub Perl ForceModel Eval register
{
  my @argStr;
  push @argStr, "ArgumentType = Output; Name = Status;
  ArgumentName = Status";
  my @descripArray;
  push @descripArray, "ArgumentType = Output";
```

```
  push @descripArray, "Name = Acceleration";
  push @descripArray, "RefName = LVLH";
  push @descripArray, "ArgumentName = accel";
  push @argStr, \@descripArray;
  push @argStr, "ArgumentType = Input; Name = Velocity;
  RefName =
Inertial;
    ArgumentName = Vel";
  push @argStr, "ArgumentType = Input; Name = DateUTC;
  ArgumentName =
Date";
  return @argStr;
}
```

## Computing Outputs from Inputs

The order of registration determines the order in the incoming and outgoing arrays of the requested data. For example, based on the registration given above, there are two requested inputs. When called upon to compute, the input array will then consist of three elements:

> Index 0: The calling mode, usually undefined
>
> Index 1: The velocity (an array of three doubles), referred to by the name *Vel* in the script
>
> Index 2: The UTC date (a String), referred to as *Date* in the script.

Accordingly, the output array consists of two elements:

> Index 0: Status (a String)
>
> Index 1: The acceleration (an array of three doubles) relative to the LVLH frame, referred to as *accel* in the script.

An error results if:

- an array is not returned,
- fewer outputs were returned than were requested,
- the incorrect type of data is returned for an argument (e.g., a String was returned where a Double was expected), or
- an array is returned that is too short (e.g., returning two doubles rather than the expected three).

If an error results, the script is turned off, and a message explaining the problem is sent to the message window.

NOTE: If the script function returns a String instead of an array, an error results which then will cause the script to be turned off. Additionally, the returned string will be written to the message window. Thus, the script has the ability to turn the script off and indicate a reason for doing so, simply by returning a string.

### Accessing Data on the Basis of ArgumentName

Because the order of registration implies the order of the data in the input array and the output array, it is imperative that the correct association between the argument and its index number be used. Thus, adding new arguments in the middle of the current list or even just reordering the list of descriptors in the register function will require modifications of the index numbers used in the *compute* function. To alleviate this burden somewhat, the user can choose to use auxiliary classes that hold the association between argument and index number. This allows the user to access data by name, which should result in fewer modifications being necessary in the *compute* function when the registration function is altered.

After the registration process occurs, STK makes an additional call to Perl to create two Classes (one for inputs, one for outputs) that can be used to access the data by ArgumentName (rather than using index numbers). These classes can be created and stored during the first call to compute, and then accessed by compute thereafter. The classes are accessed from the $g_Plug-inArrayInterfaceHash hash. For example, consider the function Perl_ForceModel_Eval. The *compute* function would look, like this:

```
sub Perl_ForceModel_Eval_compute
{
  # the inputs here are in the order of the requested Inputs,
  as registered
  my @stateData = @_;
  # $stateData [0] is the calling mode
  # This function ASSUMES that Perl_ForceModel_Eval_register
  will set
  # Perl_ForceModel_Eval_init to -1
  #
  if ($Perl_ForceModel_Eval_init < 0)
  {
    $Perl_ForceModel_Eval_init = 1;
    # The following hashes have been created automatically
    after this script
    # has registered its inputs and outputs.
    #
    # Each hash contains information about the arguments for
    this script.
    # The hashes have been created as a user convenience, for
    those users
```

```
        wanting to know, during the running of the script, what
        the inputs and
        # outputs are. In many cases, the script writer doesn't
        care, in which
case
        # this entire if-block is unneeded and can be removed.
        $Perl_ForceModel_Eval_Inputs =
          g_Plug-inArrayInterfaceHash{'Perl_ForceModel_Eval'}
          {'Inputs'};
        $Perl_ForceModel_Eval_Outputs =
          $g_Plug-inArrayInterfaceHash{'Perl_ForceModel_Eval'}
          {'Outputs'};
        # comment out the line below if you don't want to see the
        inputs and
outputs
        # each time the script is run
        # Perl_ForceModel_Eval_showArgs ();
}
    # continue with rest of script
    # compute the acceleration: here it is a "reverse" drag,
    # being proportional to the inertial speed
    my @velArray = @{$stateData[$Perl_ForceModel_Eval_Inputs-
    >getArgument('Vel')]};
    my $factor = 0.000001;
    my $cbiSpeed = sqrt($velArray[0]*$velArray[0]
        +$velArray [1] *$velArray [1] +$velArray [2]
        *$velArray [2]);
    # accel with be the acceleration in the CbiLVLH frame
    my @accel;
    push @accel, 0.0;                  # x component: radial
    push @accel, $factor*$cbiSpeed;  # y-component: inTrack
    push @accel, 0.0;                  # z-component: crossTrack
    # this defines the return array
    my @retArray = ();
    $retArray[$Perl_ForceModel_Eval_Outputs-
    >getArgument('accel')] = \@accel;
    $retArray[$Perl_ForceModel_Eval_Outputs-
    >getArgument('Status')] = "Okay";
    return @returnArray;
}
sub Perl_ForceModel_Eval_showArgs
{
    my @argStrArray;
    STKUtil::printOut "Doing Perl_ForceModel_Eval_compute_
    init\n";
    @argStrArray = ();
    push @argStrArray, $Perl_ForceModel_Eval_Inputs->
              {'FunctionName'}->{'Name'}. "Inputs \n";
    # the first arg on input is the calling mode
    push @argStrArray, "0: this is the calling mode\n";
```

```
my @args = $Perl_ForceModel_Eval_Inputs->getArgumentArray ();
# to see description args
my $index, $descrip;
foreach $arg (@args)
{
    ($index, $descrip) = $Perl_ForceModel_Eval_Inputs->
    getArgument ($arg);
    push @argStrArray, "$index: $arg = $descrip\n";
}
STKUtil::printOut @argStrArray;
@argStrArray = ();
push @argStrArray, $Perl_ForceModel_Eval_Outputs->
            {'FunctionName,}->{'Name'}. "Outputs \n";
my @args = $Perl_ForceModel_Eval_Outputs->
getArgumentArray();
# to see description args
my $index, $descrip;
foreach $arg (@args)
{
    ($index, $descrip) = $Perl_ForceModel_Eval_Outputs->
    getArgument($arg);
    push @argStrArray, "$index: $arg = $descrip\n";
}
STKUtil::printOut @argStrArray;
}
```

The variables $Perl_ForceModel_Eval_init, $Perl_ForceModel_Eval_Inputs, and $Perl_ForceModel_Eval_Outputs are global variables and must be declared outside of any function. During Perl_ForceModel_Eval_register, Perl_ForceModel_Eval_init should be set to –1. The first time Perl_ForceModel_Eval_compute is called, the if-block is executed, causing two classes to be created: (Perl_ForceModel_Eval_Inputs and Perl_ForceModel_Eval_Outputs). Because they have been declared global, they are remembered for the next call to compute.

Notice that the inputs are accessed from stateData using the class Perl_ForceModel_Eval_Inputs and the ArgumentName that was registered. Similarly, the outputs are assigned into the returned array using the class Perl_ForceModel_Eval_Outputs and the ArgumentName that was registered. The classes provide a mechanism in the script to access the inputs and outputs by name, rather by index number. Of course, they do this by mapping the ArgumentName to an index value:

| Name | Index Value |
|---|---|
| $Perl_ForceModel_Eval_Inputs->getArgument{'Vel'} | 1 |
| $Perl_ForceModel_Eval_Inputs->getArgument{'Date'} | 2 |
| $Perl_ForceModel_Eval_Outputs->getArgument{'Status'} | 0 |
| $Perl_ForceModel_Eval_Outputs->getArgument{'accel'} | 1 |

(Remember, on input the argument at index 0 is the calling mode.)

> **CAUTION:** The use of hashes to translate ArgumentName to index value as described above does affect performance: the code may execute up to twice as fast when the hashes are not used and the index values are used directly (this contrasts with VBScript where no performance penalty arises).

> **NOTE:** A matrix is passed in Perl as a single array, containing the rows of the matrix in order, not in the form of an array of references to arrays or hash.

## Astrogator Plug-in Points

The following sections describe Plug-in points related to the Astrogator module.

> **NOTE:** Additional plug-in points utilizing Microsoft COM technologies are available for Astrogator. These plug-in points include HPOP Force Model plug-ins, Search Profile plug-ins, Engine Model plug-ins, and Attitude Controller plug-ins. Documentation on these interfaces can be found in the Programming Interface Help in the STK Software Development Kit->Extend STK through Plug-ins section. Examples of these plug-ins in multiple programming languages can be found under CodeSamples\Extend in the STK install directory.

### Custom Functions

Many of the Astrogator Plug-in points are implemented using Custom Functions. A Custom Function is an Astrogator Component in the Component Browser that calls a script function defined in a file. There are four types of Custom Functions: MATLAB, VBScript, Python, and Perl. A Custom Function has only one attribute: the filename containing the function to be called. Each time a Custom Function is initialized, STK checks the timestamp of the file. If the file has been edited since its last loading, then it is automatically reloaded; otherwise, reloading is unnecessary and is not performed.

### Calculation Objects

Calc Objects can utilize scripting through the use of:

1. Inline functions, and
2. Custom Functions.

Each of these types is found in the Scripts folder within the Calculation Objects folder in the Astrogator Component Browser.

## Inline Functions

Inline functions are used for very simple computations that are not presently available as Calculation Objects in the Component Browser. They have three attributes (see Table 3).

There are four types of inline script functions available: MATLAB, VBScript, JavaScript, and Perl. When an inline Calculation Object is used, STK creates a function with the inputs listed in *CalcArguments*, returning the value computed by the *InlineFunc*, with a name derived from the CalculationObject name itself. Example:

```
CalcObject Name:  Incl cos Raan
CalcArguments:    Inclination, RAAN
InlineFunc:       Inclination*cos(RAAN)
UnitDimension:    AngleUnit
```

An inline function named CalcObject_Incl_cos_Raan is created, with Inclination and RAAN as inputs. The function returns the value Inclination*cos(RAAN), which is interpreted as an Angle. The *InlineFunc* expression must use the correct syntax of the scripting language but see the exception concerning Perl).

The MATLAB Calculation Object uses the MATLAB inline function to define the function:

```
CalcObject_Incl_cos_Raan = inline ('Inclination*cos (RAAN)',
'Inclination', 'RAAN');
```

The VBScript Calculation Object creates the function as follows:

```
Function CalcObject_Incl_cos_Raan (Inclination, RAAN)
  Dim value
  value = Inclination*cos (RAAN)
  CalcObject_Incl_cos_Raan = value
End Function
```

TABLE 3

Attributes of Inline Functions

| Attribute | Description |
| --- | --- |
| CalcArguments | A list of Calculation Objects which are to be inputs to the functions |
| InlineFunc | The body of the function that uses the inputs to compute one double value |
| UnitDimension | The dimension that the resulting value is declared to have |

The JavaScript Calculation Object creates the function as follows:

```
function CalcObject_Incl_cos_Raan (Inclination, RAAN)
{
    return eval ("Inclination*cos (RAAN)");
}
```

Perl defines functions differently from MATLAB and VBScript. The Perl Calculation Object equivalent function is defined as follows:

```
sub CalcObject_Incl_cos_Raan
{
    return $_[0] *cos ($_[1]);
}
```

Note that there is no method for declaring the arguments as Inclination and RAAN; they have been replaced by $_[0] and $_[1], respectively, based on the knowledge that the order of the arguments is Inclination followed by RAAN. Users may enter the InlineFunc value in proper Perl notation as shown above (complete with the scalar prefix $) but then are responsible for translating each argument to the function by the proper symbol $_[n], where n is the index of the argument (beginning with 0 in Perl).

Because many users may not be familiar with Perl syntax and yet want to use Perl, the Perl Calculation Object will create the proper Perl syntax automatically from an *InlineFunc* written using the names—thus, users may enter Inclination*cos (RAAN) as the *InlineFunc* even for Perl. When utilizing this behavior, be careful not to mix the two approaches: $Inclination*cos ($RAAN) will result in a syntax error.

> **CAUTION**: Because the name of the Calculation Object is used in the name of the function being defined, it must be a valid function name. Spaces in the name are allowed because STK will substitute underscores (i.e.,'_'); MATLAB, VBScript, and Perl each allows underscores in function names. Symbols such as '(' or '*' etc. are not allowed, however.

## Custom Function Calc Objects

Calculation Objects based upon Custom Functions are available for use with computations that are lengthy or involved. Unlike the inline functions, Custom Function Calc Objects require users to write a script stored in a file. See the documentation for Custom Functions.

Custom Function Calc Objects have two Entry Points that can be utilized. Use of any Entry Point is optional, though the value of the calc Object will be 0.0 if the Eval Entry Point is not used (see Table 4).

The Eval Entry Point has the following available inputs and outputs, as shown in Table 5.

**TABLE 4**

Entry Points for Custom Function Calc Objects

| Entry Point | Event | Intended Use |
|---|---|---|
| Eval | Called at every Calc Object evaluation | To return the value of the Calc Object |
| Reset | Called every time script is reinitialized (before computing, before each segment runs, and before reporting) | To initialize any variables in preparation for Eval to be used |

**TABLE 5**

Custom Function Calc Objects—Eval Entry Point Arguments and Keywords

| Name | ArgumentType | Type | Additional Keywords | Additional Keyword Options |
|---|---|---|---|---|
| Value | Output | — | — | — |
| <name> | Input | CalcObject | — | — |
| <name> | Input | GatorVector | RefName | <refAxes> |
| Epoch | Input | — | — | — |
| DateUTC | Input | — | — | — |

In the above table, <name> indicates the name of an object of the given Type from the Astrogator Component Browser, and <refAxes> indicates the name of Axes from the Astrogator Component Browser.

The Reset Entry point has the same inputs as the Eval Entry point, but no outputs.

## Custom Engine Models

The Custom Engine Models has five Entry Points that may be utilized. Use of any Entry Point is optional (see Table 6).

## Inputs and Outputs

The Pre-Propagation and Post-Propagation Entry Points have no inputs or outputs. (The script function signature, however, must still allow for one argument).

The Update Entry point has the following available inputs and outputs, as shown in Table 7.

In the above table, <name> indicates the name of an object of the given Type from the Astrogator Component Browser, and <refAxes> indicates the name of an Axes from the Astrogator Component Browser.

The SegmentStart Entry Point has the same inputs that are available in the Update Entry Point, but has no outputs.

The Eval Entry point has the following available inputs and outputs (see Table 8).

**TABLE 6**

Entry Points for Custom Engine Models

| Entry Point | Event | Intended Use |
|---|---|---|
| Post-Propagation | Called before any propagation begins | To perform any necessary preprocessing before the Update or Eval scripts are called. Examples: load a file used the Eval script; open a socket |
| Post-Propagation | Called after all propagation ends | To perform any necessary cleanup operation. Examples: let go of a file handle; close a socket |
| SegmentStart | Called when a new segment starts | To perform any necessary initialization |
| Update | Called at the beginning of every integration step. Also called on the last state of the segment | To update a parameter in a discontinuous way. No discontinuities in parameter values should be made during an integration step; i.e., none should be made using Eval. Also, to update a parameter in a permanent manner. |
| Eval | Called at every force model evaluation | To return engine thrust, mass flow rate, and optionally specific impulse |

**TABLE 7**

Custom Engine Models—Update Entry Point Arguments and Keywords

| Name | ArgumentType | Type | Additional Keywords | Additional Keyword Options |
|---|---|---|---|---|
| Status | Output | — | — | — |
| <name> | Input | CalcObject | — | — |
| <name> | Input | GatorVector | RefName | <ref-Axes> |
| Status | Input | — | — | — |
| DateUTC | Input | — | — | — |
| CbName | Input | — | — | — |
| Epoch | Input | — | — | — |
| Mu | Input | — | — | — |
| TotalMass | Input | — | — | — |
| DryMass | Input | — | — | — |
| FuelMass | Input | — | — | — |
| CD | Input | — | — | — |
| CR | Input | — | — | |
| DragArea | Input | — | — | — |
| SRPArea | Input | — | — | — |
| Position | Input | — | RefName | Inertial |
| | | | RefName | Fixed |
| Velocity | Input | — | RefName | Inertial |
| | | | RefName | Fixed |

**TABLE 8**

Custom Engine Models—Eval Entry Point Arguments and Keywords

| Name | ArgumentType | Type | Additional Keywords | Additional Keyword Options |
|---|---|---|---|---|
| Status | Output | — | — | — |
| Thrust | Output | — | — | — |
| Isp | Output | — | — | — |
| MassFlowRate | Output | — | — | — |
| <name> | Input | CalcObject | — | — |
| <name> | Input | GatorVector | RefName | <refAxes> |
| Status | Input | — | — | — |
| DateUTC | Input | — | — | — |
| CbName | Input | — | — | — |
| Epoch | Input | — | — | — |
| Mu | Input | — | — | — |
| TotalMass | Input | — | — | — |
| DryMass | Input | — | — | — |
| FuelMass | Input | — | — | — |
| CD | Input | — | — | — |
| CR | Input | — | — | — |
| DragArea | Input | — | — | — |
| SRPArea | Input | — | — | — |
| Position | Input | — | RefName | Inertial |
| | | | RefName | Fixed |
| Velocity | Input | — | RefName | Inertial |
| | | | RefName | Fixed |

In the above table, <name> indicates the name of an object of the given Type from the Astrogator Component Browser, and <refAxes> indicates the name of an Axes from the Astrogator Component Browser.

## Propagators

All Propagators have five Entry Points that may be utilized. Use of any Entry Point is optional (see Table 9).

## Inputs and Outputs

The Pre-Propagation and Post-Propagation Entry Points have no inputs or outputs. (The script function signature, however, must still allow for one argument).

The Update Entry point has the following available inputs and outputs, as shown in Table 10.

**TABLE 9**

Entry Points for Propagators

| Entry Point | Event | Intended Use |
|---|---|---|
| Pre-Propagation | Called before any propagation begins | To perform any necessary preprocessing before the Update or Eval scripts are called. Examples: load a file used in the Eval script; open a socket |
| Post-Propagation | Called after all propagation ends | To perform any necessary cleanup operation. Examples: let go of a file handle; close a socket |
| SegmentStart | Called when a new segment starts | To perform any necessary initialization. |
| Update | Called at the beginning of every integration step. Also called on the last state of the segment | To update a parameter in a discontinuous way. No discontinuities in parameter values should be made during an integration step; i.e., none should be made using Eval. Also, to update a parameter in a permanent manner, so that subsequent computations (including subsequent segments if applicable) utilize the updated value (unless overridden in Eval) |
| Eval | Called at every force model evaluation | To return an additional acceleration; and/or to set certain parameter values (DragArea, CD, density, etc.) that are to be used during this force model computation—parameter changes are not permanent and are not remembered |

In the above table, <name> indicates the name of an object of the given Type from the Astrogator Component Browser, and <refAxes> indicates the name of an Axes from the Astrogator Component Browser.

> **NOTE:** When CD, CR, DragArea, SRPArea, DryMass, and/or FuelMass are set as outputs, the values that are assigned permanently affect the satellite's physical properties as viewed on the GUI or reported in the MCS Summary. Thus, assigning values for these during Update will affect the force model computation, unless overridden by the Eval Entry Point. In addition, subsequent MCS segments will use these updated values as well.

The SegmentStart Entry Point has the same inputs that are available in the Update Entry Point, but has no outputs.

The Eval Entry point has the following available inputs and outputs (see Table 11).

In the above table, <name> indicates the name of an object of the given Type from the Astrogator Component Browser, and <refAxes> indicates the name of an Axes from the Astrogator Component Browser.

**TABLE 10**

Propagators—Update Entry Point Arguments and Keywords

| Name | ArgumentType | Type | Additional Keywords | Additional Keyword Options |
|---|---|---|---|---|
| Status | Output | — | — | — |
| CD | Output | — | — | — |
| CR | Output | — | — | — |
| DragArea | Output | — | — | — |
| SRPArea | Output | — | — | — |
| DryMass | Output | — | — | — |
| FuelMass | Output | — | — | — |
| <name> | Input | CalcObject | — | — |
| <name> | Input | GatorVector | RefName | <refAxes> |
| Status | Input | — | — | — |
| DateUTC | Input | — | — | — |
| CbName | Input | — | — | — |
| Epoch | Input | — | — | — |
| Mu | Input | — | — | — |
| TotalMass | Input | — | — | — |
| DryMass | Input | — | — | — |
| FuelMass | Input | — | — | — |
| CD | Input | — | — | — |
| CR | Input | — | — | — |
| DragArea | Input | — | — | — |
| SRPArea | Input | — | — | — |
| Position | Input | — | RefName | Inertial |
| | | | RefName | Fixed |
| Velocity | Input | — | RefName | Inertial |
| | | | RefName | Fixed |

NOTE: When CD, CR, DragArea, SRPArea, Density, and/or SolarIntensity are set as outputs, the values that are assigned permanently affect the satellite's physical properties as viewed on the GUI or reported in the MCS Summary. Thus, assigning values for these during Update will affect the force model computation, unless overridden by the Eval Entry Point. In addition, subsequent MCS segments will use these updated values as well.

## Information on Arguments

### *Representation and Units*

In the following table, arguments are identified by the Name element and, where necessary, the Type element. For each argument, the Representation (essentially, the data type) and Units are given (see Table 12).

**TABLE 11**

Propagators—Eval Entry Point Arguments and Keywords

| Name | ArgumentType | Type | Additional Keywords | Additional Keyword Options |
|---|---|---|---|---|
| Status | Output | — | — | — |
| CD | Output | — | | |
| CR | Output | — | — | — |
| DragArea | Output | — | — | — |
| SRPArea | Output | — | — | — |
| Density | Output | — | — | — |
| SolarIntensity | Output | | — | — |
| Acceleration | Output | — | RefName | Inertial |
| | | | RefName | Fixed |
| | | | RefName | CbiVNC |
| | | | RefName | CbiLVLH |
| <name> | Input | CalcObject | — | — |
| <name> | Input | GatorVector | RefName | <refAxes> |
| Status | Input | — | — | — |
| DateUTC | Input | — | — | — |
| CbName | Input | — | — | — |
| Epoch | Input | — | — | — |
| Mu | Input | — | — | — |
| TotalMass | Input | — | — | — |
| DryMass | Input | — | — | — |
| FuelMass | Input | — | — | — |
| CD | Input | — | — | — |
| CR | Input | — | — | — |
| DragArea | Input | — | — | — |
| SRPArea | Input | — | — | — |
| Position | Input | — | RefName | Inertial |
| | | | RefName | Fixed |
| Velocity | Input | — | RefName | Inertial |
| | | | RefName | Fixed |
| Density | Input | — | — | — |
| SolarIntensity | Input | — | — | — |
| AtmTemperature | Input | — | — | — |
| AtmPressure | Input | — | — | — |
| DragAltitude | Input | — | — | — |
| Longitude | Input | — | — | — |
| Latitude | Input | — | — | — |
| SatToSunVector | Input | — | SunPosType | True Apparent ApparentToTrueCb |

**TABLE 12**

Entry Point Representation and Units

| Name (Type) | Representation | Units |
| --- | --- | --- |
| <name> (CalcObject) | Double | based on CalcObject |
| <name> (GatorVector) | Double:3 | STK internal |
| Acceleration | Double:3 | m/sec$^2$ |
| AtmPressure | Double | N/m$^2$ |
| AtmTemperature | Double | Kelvin |
| CbName | String | — |
| CD | Double | unitless |
| CR | Double | unitless |
| DateUTC | String | DDD/YYYY HH:MM:SS.ss |
| Density | Double | kg/m$^3$ |
| DragAltitude | Double | m |
| DragArea | Double | M$^2$ |
| DryMass | Double | kg |
| Epoch | Double | epoch secs |
| FuelMass | Double | kg |
| Isp | Double | secs |
| Latitude | Double | rad |
| Longitude | Double | rad |
| MassFlowRate | Double | kg/sec |
| Mu | Double | M$^3$/sec$^2$ |
| Position | Double:3 | m |
| SatToSunVector | Double:3 | m |
| SolarIntensity | Double | unitless |
| SRPArea | Double | M$^2$ |
| Status | String | – |
| Thrust | Double | N |
| TotalMass | Double | kg |
| Velocity | Double:3 | m/sec |

In the above table, <name> indicates the name of an object of the given Type from the Astrogator Component Browser.

### Glossary

The following table explains some of the terms used in defining arguments to Entry Points for Astrogator Plug-in Points (see Table 13).

**TABLE 13**

Astrogator Glossary

| Parameter | Definition |
| --- | --- |
| Acceleration | The additional acceleration to add to the acceleration computed according to the Propagator's Force Model settings. |
| AtmPressure | The pressure as indicated by the Atmospheric Model, if available. If not, a negative number is given. |
| AtmTemperature | The temperature as indicated by the Atmospheric Model, if available. If not, a negative number is given. |
| CbName | The name of the central body that acts as the reference frame source for the Position and Velocity inputs. |
| DateUTC | The UTC date equivalent to Epoch, given in the format DDD/YYYY HH:MM:SS.ss. One can extract from this string the calendar date if such knowledge is required in the script. |
| DragAltitude | The altitude that was used in the drag model to compute the Density. |
| Epoch | The time, in epoch seconds. |
| FuelMass | During Eval, this is part of the integrator's state data. |
| Latitude | The detic latitude with reference to the central body indicated by CbName, corresponding to the Position at this Epoch. |
| Longitude | The longitude corresponding to the Position at this Epoch. |
| Mu | The gravitational parameter associated with the central body named by CbName. |
| Position | The position at this Epoch, in the frame indicated, with reference to the central body indicated by CbName. During Eval, this is part of the integrator's state data. |
| SatToSunVector | The position of the Sun, in the Inertial frame of the central body indicated by CbName, as used in SRP computations. |
| SolarIntensity | A measure of the obstruction of the solar disk, as viewed from Position. A value of 1.0 indicates no obstruction, and value of 0.0 indicates full obstruction (i.e., umbra), and a value between 0.0 and 1.0 is partial obstruction (i.e., penumbra). |
| Status | On input, value is set to OK. As an output, its value is checked against the words "Error", "Stop", and "Cancel" in a case-insensitive manner. If one of these was returned as an output, then the script is turned off and a message (indicating which string was found) is returned to the message window. This provides a mechanism within the script for turning it off. |
| TotalMass | The sum of DryMass plus FuelMass. |
| Velocity | The velocity at this Epoch, in the frame indicated, with reference to the central body indicated by CbName. During Eval, this is part of the integrator's state data. |

## Vector Geometry Tool Plug-in Points

The Vector Geometry Tool allows users to create new Axes, Vectors, Angles, etc. whose definitions are based upon other components of the tool.

Additionally, it provides for two Plug-in Points: Custom Script Vectors and Custom Script Axes. These are used to create Vector and Axes objects that cannot be created using the existing Vector Geometry Tool methods. The available inputs consist of any object within the Vector Geometry Tool and a special visibility input that can perform line-of-sight and field-of-view computations. Vector Geometry Tool objects can be accessed as inputs for other Plug-in Points as well.

Custom Vectors and Custom Axes are initialized when loaded from a file or created within STK. They are not automatically reinitialized when the file is subsequently reedited; after reediting, the user must force the script to be reloaded using the GUI by choosing to Modify the Custom Vector or Custom Axes and hitting the Reload button.

### Script Source File Location

Script files that are to be used with Custom Vectors and Custom Axes must be located in a proper directory. Custom Vector scripts must be located in one of the following directories:

(i) <STK install folder>/STKData/Scripting/VectorTool/Vector

(ii) <STK user area>/Scripting/VectorTool/Vector

(iii) <scenario folder>/Scripting/VectorTool/Vector

where <STK install folder> refers to the directory path that is the parent of STKData, <STK user area> refers to the user's configuration directory, and <scenario folder> refers to the scenario directory. Likewise, Custom Axes scripts must be located in one of the following:

(i) <STK install folder>/STKData/Scripting/VectorTool/Axes

(ii) <STK user area>/Scripting/VectorTool/Axes

(iii) <scenario folder>/Scripting/VectorTool/Axes

> **WARNING:** Do not place your scripts directly in the Scenario folder. Doing so may cause errors with your script.

All users running the same installed version of STK share the same <STK install folder> directory and thus share all the Vector and Axes scripts stored there; all Vector and Axes scripts stored relative to <STK user area> are available for use for a user of any scenario; all Vector and Axes scripts stored relative to <scenario folder> are only available for that scenario.

> **NOTE:** If the user changes the scenario directory (e.g., by loading a new scenario or saving the current scenario in a new location), then the Custom Vector/Axes scripts will be loaded using this new scenario

directory, not the old one. In particular, the user will need to create the appropriate directories if the user saves a scenario into a new location.

A script file is not automatically reloaded if edited; you must use the reload button on the GUI to force it to be reloaded. The script is called whenever the value of the object (its location, orientation, rate, etc.) is needed.

## Custom Vector

The Custom Vector has the following outputs, as shown in Table 14.

The Vector refers to the $x$, $y$, and $z$ components of the Vector with respect to the Reference Axes set for the Custom Vector; if the derivative of the vector as observed in the Reference Axes frame is known, then the script can output VectorRate. If VectorRate is not used, then the Custom Vector uses a numerical approximation to compute its derivative when requested.

## Custom Axes

The Custom Axes have the following outputs (see Table 15).

Quat refers to the quaternion describing the orientation of the Custom Axes with respect to the Reference Axes that have been chosen. The first three elements of Quat represent the vector portion of the quaternion, while the fourth element represents the scalar part. The script may output the angular velocity of the frame as observed by the Reference Axes, if this is known, by outputing AngVel as well. If AngVel is not used, then the Custom Axes uses a numerical approximation to compute its angular velocity when requested.

**TABLE 14**

Custom Vector Outputs

| Keyword | ArgumentType | Name | Representation |
|---------|--------------|------|----------------|
| Value | Output | Vector | Double:3 |
| Value | Output | VectorRate | Double:3 |

**TABLE 15**

Custom Axes Outputs

| Keyword | ArgumentType | Name | Representation |
|---------|--------------|------|----------------|
| Value | Output | Quat | Double:4 |
| Value | Output | AngVel | Double:3 |

## Available Inputs

All objects within the Vector Geometry Tool are available as inputs to Custom Script Vector and Custom Script Axes, as well as the time in epoch seconds. The descriptors for each type of input are shown below.

### *Epoch*

Epoch has a single descriptor, Name, as shown in Table 16.:
    The argument is a Double in units of epoch seconds.

### *Vector*

The descriptors for the Vector input type are as follows (see Table 17).

Type, Name, and RefName are all required; Source, RefSource, and Derivative are optional. If the Source keyword is not specified, then a default value is used. For Custom Vectors and Custom Axes, the default value is the STK path for the object that owns the Custom Vector/Axes. The RefSource keyword works similarly. The combination of Name and Source uniquely identifies the requested Vector; the vector is then resolved into components as observed in the axes specified by RefName and RefSource when its value is computed and sent to the script. If Derivative is not set or is not Yes, then just the value is given as Double:3; if Derivative has been set to Yes, then the vector and its vector derivative are computed and the values are given as Double:6 (the first three specifying the vector, the last three its derivative).

**TABLE 16**

Epoch Descriptor

| Keyword | Value |
|---------|-------|
| Name | Epoch |

**TABLE 17**

Vector Descriptors

| Keyword | Value |
|---------|-------|
| Type | Vector |
| Name | Name of the requested vector |
| Source | STK path for the object where the vector of given Name resides. To reference the parent or grandparent object, use <MyParent> or <MyGrandParent> (including the opening and closing brackets). |
| RefName | Name of requested reference axes |
| RefSource | STK path for the object where the axes of given RefName reside. To reference the parent or grandparent object, use <MyParent> or <MyGrandParent> (including the opening and closing brackets). |
| Derivative | Yes or No |

The units of the components depend on the units associated with the vector; the derivative modifies the unit to be per second.

### Axes

The descriptors for the Axes input type are as follows (see Table 18).

Type, Name, and RefName are all required; Source, RefSource, and Derivative are optional. If the Source keyword is not specified, then a default value is used. For Custom Vectors and Custom Axes, the default value is the STK path for the object that owns the Custom Vector/Axes. The RefSource keyword works similarly. The combination of Name and Source uniquely identifies the requested Axes; the value of the axes is then the quaternion relating the orientation of these requested axes as observed in the axes specified by RefName and RefSource. If Derivative is not set or is not Yes, then just the quaternion is given as Double:4; if Derivative has been set to Yes, then the quaternion and the angular velocity are computed and the values are given as Double:7 (the first four specifying the quaternion, the last three its angular velocity). The angular velocity is in units of per second.

### Angle

The descriptors for the Angle input type are as follows (see Table 19).

**TABLE 18**

Axes Descriptors

| Keyword | Value |
| --- | --- |
| Type | Axes |
| Name | Name of the requested Axes |
| Source | STK path for the object where the axes of given Name reside. To reference the parent or grandparent object, use <MyParent> or <MyGrandParent> (including the opening and closing brackets). |
| RefName | Name of requested reference axes |
| RefSource | STK path for the object where the axes of given RefName reside. To reference the parent or grandparent object, use <MyParent> or <MyGrandParent> (including the opening and closing brackets). |
| Derivative | Yes or No |

**TABLE 19**

Angle Descriptors

| Keyword | Value |
| --- | --- |
| Type | Angle |
| Name | Name of the requested Angle |
| Source | STK path for the object where the angle of given Name resides |
| Derivative | Yes or No |

Type and Name are each required; Source and Derivative are optional. If the Source keyword is not specified, then a default value is used. For Custom Vectors and Custom Axes, the default value is the STK path for the object that owns the Custom Vector/Axes.

The combination of Name and Source uniquely identifies the requested Angle. If Derivative is not set or is not Yes, then just the angle is given as Double; if Derivative has been set to Yes, then the value and its derivative are computed and the values are given as Double:2. The angle is in radians; angle rate in radians per second.

### Point

The descriptors for the Point input type are as follows (see Table 20).

Type, Name, and RefName are all required; Source, RefSource, and Derivative are optional. If the Source keyword is not specified, then a default value is used. For Custom Vectors and Custom Axes, the default value is the STK path for the object that owns the Custom Vector/Axes. The RefSource keyword works similarly. The combination of Name and Source uniquely identifies the requested Point; the location of that point is then resolved into components as observed in the coordinate system specified by RefName and RefSource when its value is computed and sent to the script. If Derivative is not set or is not Yes, then just the location is given as Double:3; if Derivative has been set to Yes, then the location and its derivative are computed and the values are given as Double:6 (the first three specifying the location, the last three its derivative). The location is in meters, its derivative in meters per second.

### CrdnSystem

The descriptors for the CrdnSystem input type are as follows (see Table 21).

**TABLE 20**

Point Descriptors

| Keyword | Value |
| --- | --- |
| Type | Point |
| Name | Name of the requested Point |
| Source | STK path for the object where the point of given Name resides. To reference the parent or grandparent object, use <MyParent> or <MyGrandParent> (including the opening and closing brackets). |
| RefName | Name of requested reference coordinate system |
| RefSource | STK path for the object where the coordinate system of given RefName resides. To reference the parent or grandparent object, use <MyParent> or MyGrandParent> (including the opening and closing rackets). |
| Derivative | Yes or No |

**TABLE 21**

CrdnSystem Descriptors

| Keyword | Value |
|---|---|
| Type | CrdnSystem |
| Name | Name of the requested CrdnSystem |
| Source | STK path for the object where the coordinate system of given Name resides. To reference the parent or grandparent object, use <MyParent> or <MyGrandParent> (including the opening and closing brackets). |
| RefName | Name of requested reference coordinate system |
| RefSource | STK path for the object where the coordinate system of given RefName resides. To reference the parent or grandparent object, use <MyParent> or <MyGrandParent> (including the opening and closing brackets). |
| Derivative | Yes or No |

Type, Name, and RefName are all required; Source, RefSource, and Derivative are optional. If the Source keyword is not specified, then a default value is used. For Custom Vectors and Custom Axes, the default value is the STK path for the object that owns the Custom Vector/Axes. The RefSource keyword works similarly. The combination of Name and Source uniquely identifies the requested CrdnSystem. The CrdnSystem provides the location of the origin and orientation of its axes relative to the coordinate system specified by RefName and RefSource. If Derivative is not set or is not Yes, then the script is given Double:7 where the location is the first three elements, and the quaternion is the last four. If Derivative has been set to Yes, then the script is given Double:13 where the first three elements are location, the next three elements its derivative, the next four the quaternion, and the last three the angular velocity. The location is in meters, its derivative in meters per second.

### *Plane*

The descriptors for the Plane input type are as follows (see Table 22).

Type, Name, and RefName are all required; Source, RefSource, RefType, and Derivative are optional. If the Source keyword is not specified, then a default value is used. For Custom Vectors and Custom Axes, the default value is the STK path for the object that owns the Custom Vector/Axes. The RefSource keyword works similarly. The combination of Name and Source uniquely identifies the requested plane; the plane is then resolved into components as observed in the axes or coordinate system specified by RefName, RefSource, and RefType when its value is computed and sent to the script. The returns for the four possible combinations of RefType and Derivative values are shown in Table 23.

The location is in meters, its derivative in meters per second.

**TABLE 22**

Plane Descriptors

| Keyword | Value |
|---------|-------|
| Type | Plane |
| Name | Name of the requested plane |
| Source | STK path for the object where the plane of given Name resides. To reference the parent or grandparent object, use MyParent> or <MyGrandParent> (including the opening and closing brackets). |
| RefName | Name of requested reference axes or coordinate system see RefType, below. |
| RefSource | STK path for the object where the axes or coordinate system of given RefName reside (see RefType, below). To reference the parent or grandparent object, use MyParent> or <MyGrandParent> (including the opening and closing brackets). |
| RefType | Set to Axes or CrdnSystem. Defaults to Axes. |
| Derivative | Yes or No. Defaults to No. |

**TABLE 23**

Returns from Plane Input Type

| RefType | Derivative | Returns |
|---------|-----------|---------|
| Axes | No | The plane's Axis1 and Axis2 are resolved into Axes specified by RefName. The output is 6 doubles: first 3 for Axis1, next 3 for Axis2. |
| Axes | Yes | The plane's Axis1 and Axis2 are resolved into Axes specified by RefName. The output is 12 doubles: first 3 for Axis1, next 3 for Axis2, next 3 for velocity of Axis1, next 3 for velocity of Axis2. |
| CrdnSystem | No | The plane's Axis1 and Axis2 are resolved into Axes owned by the CrdnSystem specified by RefName. The plane's reference point position and velocity are found relative to this same CrdnSystem. The output is 9 doubles: first 3 for Axis1, next 3 for Axis2, last 3 for position of reference point. |
| CrdnSystem | Yes | The plane's Axis1 and Axis2 are resolved into Axes owned by the CrdnSystem specified by RefName. The plane's reference point position and velocity are found relative to this same CrdnSystem. The output is 18 doubles: first 3 for Axis1i, next 3 for Axis2, next 3 for position of reference point, next 3 for velocity of Axis1, next 3 for velocity of Axis2, last 3 for velocity of reference point. |

### UserInfo

UserInfo has a single descriptor, Type, shown in Table 24.
   UserInfo returns the parent object, e.g., Satellite/Satellite1.

### Visibility

The descriptors for the Visibility input type are as follows (see Table 25).

**TABLE 24**

UserInfo Descriptor

| Keyword | Value |
| --- | --- |
| Type | UserInfo |

**TABLE 25**

Visibility Descriptors

| Keyword | Value |
| --- | --- |
| Type | Visibility |
| Name | Name of a Point |
| Source | STK path for the object where the Point of given Name resides |
| TargetName | Name of a Vector |
| TargetSource | STK path for the object where the Vector of given TargetName resides |
| Background | Space or CentralBody |
| FOV | Yes or No |

Type and TargetName are required; Name, Source, RefSource, Background, and FOV are optional. If the Name keyword is not given, then the Point named "Center" is used. If the Source keyword is not specified, then a default value is used. For Custom Vectors and Custom Axes, the default value is the STK path for the object that owns the Custom Vector/Axes. The TargetSource keyword works similarly.

This input provides a unitless Double value that represents a measure of visibility at the location of the Point along the direction of the Vector. The value is negative if there is no visibility, positive if there is visibility. If the Vector does not have units of distance, then the computation considers the line to infinity from the Point along the Vector direction: if the line intersects the central body associated with the Source, then a negative number results (no visibility); if the line does not intersect it, then a positive number results (visibility).

If the Vector has units of distance, the visibility is computed only within that distance from the Point. For such cases, the user can further restrict when visibility is said to occur by specifying the Background keyword. Consider the case where the Target is visible from the Point along the Vector direction and within the Vector's distance: if Background is set to Space then this case will indicate visibility only if the central body of the Source is not behind the Target along the Vector direction; if Background is set to CentralBody, then this case will indicate visibility only if the central body of the source is behind the Target.

Moreover, if the Vector is a displacement vector (and thus necessarily has distance units), not only is the central body associated with the Source checked against, but so are the central bodies associated with the source objects for the origin and destination Points of the displacement vector. Finally, if the Source is a Sensor, one can set FOV to Yes and visibility will not only be measured along the Vector direction, but also within the sensor's field of view (positive still meaning "has visibility"). Because of all the complicated options, the visibility value may not vary smoothly over time.

## Attitude Simulator Plug-in Points

Attitude Simulator is a new tool, introduced in STK version 4.3, that is available for a Satellite object. The tool provides users the ability to incorporate their own torque models (e.g., gravity gradient, aerodynamic, etc.) and momentum biases, and implement customized control laws (including static and dynamic feedback, dynamic compensators, etc.) when generating attitude trajectories. See the documentation on the Attitude Simulator for more information.

The tool numerically integrates an attitude state (represented by a quaternion) and its body angular velocity components as well as other optional variables. Attitude Simulator Plug-in Scripts may supply these variables and generally perform custom computations during numerical integration.

### Script Source File Location

The script source files must reside in the proper directory search path to be able to utilize the Plug-in Script functionality. They must be located in one of the following directories:

    (i)  &lt;STK install folder&gt;/STKData/Scripting/Attitude

    (ii)  &lt;STK user area&gt;/Scripting/Attitude

    (iii)  &lt;scenario folder&gt;/Scripting/Attitude

where &lt;STK install folder&gt; refers to the directory path that is the parent of STKData, &lt;STK user area&gt; refers to the user's configuration directory, and &lt;scenario folder&gt; refers to the scenario directory.

> **WARNING:** Do not place your scripts directly in the Scenario folder. Doing so may cause errors with your script.

### Outputs

The Attitude Simulator Plug-in Point has the following available outputs.

#### *Torque*

The torque is Double:3 and has units of N-m (see Table 26). The Torque output is used during the attitude simulation when integrating the attitude

**TABLE 26**

Torque Descriptors

| Keyword | Value |
|---|---|
| ArgumentType | Output |
| Type | Parameter |
| Name | Torque |
| BasicType | Vector |

dynamics. The attitude simulator allows many different Plug-in Scripts to be used within the same simulation: all the Torque outputs from all scripts are added together when integrating the attitude dynamics.

## *Parameter*

The Parameter output is used to share information in the current script to other Plug-in Scripts (see Table 27). Other scripts can request this script's output as an input (see below). The parameter may be a Double, integer (Int), String, Vector (Double:3), Quat (Double:4), or Matrix. If the BasicType is Matrix, then a 3×3 matrix is assumed if the Size keyword is not given. Other sizes can be specified using the Size keyword (whose value is a String containing two integers separated by a space representing the number of rows and the number of columns in the matrix).

## *Integrated Parameter*

An Integrated Parameter is a parameter that the Attitude Simulator will add to its integrator state to be integrated along with the attitude and attitude rate (see Table 28). The descriptor actually creates two Parameters that can be accessed as inputs: itself (accessed by Name) and its integral (accessed by IntegralName). The InitialCondition keyword is optional: if not given, zeroes are used. MomentumBias is a special value for IntegralName that allows user to integrate internal angular momentum due to momentum bias devices.

**TABLE 27**

Parameter Descriptors

| Keyword | Value |
|---|---|
| ArgumentType | Output |
| Type | Parameter |
| Name | Name of the parameter when accessed by this or other scripts |
| BasicType | Double, Int, String, Vector, Quat or Matrix |
| Size | String containing numberOfRows numberOfColumns |

TABLE 28

Integrated Parameter Descriptors

| Keyword | Value |
| --- | --- |
| ArgumentType | Output |
| Type | Parameter |
| Name | Name of the parameter when accessed by this or other scripts |
| BasicType | Double, Vector or Quat |
| IntegralName | Name given to the integral of this parameter |
| InitialCondition | String containing initial values of the integrated parameter separated by spaces |

> NOTE: Several Parameters, each with different Names, can all affect the same integrated state by specifying the same IntegralName (in such cases the values of all Parameters with the same IntegralName are added to create the derivative information).

## Inputs

The Attitude Simulator Plug-in Point has the following available inputs:

### *Attitude State*

The attitude state refers to the attitude being integrated within the simulator (see Table 29). The Derivative keyword is optional (default is No). If the value of Derivative is Yes, the input is Double:7 (the first four doubles being the quaternion, the last three being the angular velocity in units of 1/sec); otherwise, the input is Double:4 (the quaternion). The RefName keyword is optional: if not specified, then the vehicle's central body inertial axes are used. The RefSource keyword is optional: if not specified, it refers to the attitude simulator vehicle.

TABLE 29

Attitude State Descriptors

| Keyword | Value |
| --- | --- |
| ArgumentType | Input |
| Type | Attitude |
| Derivative | Yes or No |
| RefName | Name of axes: the attitude state will be transformed into these Axes |
| RefSource | STK path of the object that owns the RefName axes |

### Parameter

The Parameter input is used to access information in the current script from another script's Output Parameter (see Table 30).

### Integral Parameter

The Integral Parameter input is used to access the integrated parameter from the attitude simulator state (see Table 31). Note that the Name of the input must be the same as the IntegralName of some script's integrated parameter output.

### Inertia

The inertia is a 3×3 matrix of Doubles in units of kg-m$^2$ (see Table 32).

### Mass

The mass is a Double in units of kg (see Table 33).

TABLE 30

Parameter Descriptors

| Keyword | Value |
| --- | --- |
| ArgumentType | Input |
| Type | Parameter |
| Name | Name of the parameter when accessed by this or other scripts |
| Source | The filename of the script where the Parameter Name has been declared to be an Output |

TABLE 31

Integral Parameter Descriptors

| Keyword | Value |
| --- | --- |
| ArgumentType | Input |
| Type | Integral |
| Name | IntegralName of an Integrated Parameter Output |

TABLE 32

Inertia Descriptors

| Keyword | Value |
| --- | --- |
| ArgumentType | Input |
| Type | Inertia |
| Source | STK path for the object whose inertia is being requested |

**TABLE 33**

Mass Descriptors

| Keyword | Value |
|---|---|
| ArgumentType | Input |
| Type | Mass |
| Source | STK path for the object whose mass is being requested |

**TABLE 34**

Density Descriptors

| Keyword | Value |
|---|---|
| ArgumentType | Input |
| Type | Density |

### *Density*

The density is a Double in units of kg/m³ (see Table 34). The attitude simulator allows the user to configure the density modeling.

### Inputs Available from the Vector Geometry Tool

All of the inputs of the Vector Geometry Tool are available as inputs for the Attitude Simulator. In addition, those descriptors in the Vector Geometry Tool that use the keywords RefName and RefSource to refer to an Axes object can now optionally use a new keyword, RefType. Rather than specifying the RefName and RefSource to choose the references axes that the input will be transformed into, the script can specify to use RefType with the value Attitude. This designation causes the transformation to use the Attitude Simulator's state as the reference.

In addition, if you are using a Visibility argument, such that:

1. The Source of the Visibility argument is a sensor, and
2. The sensor is attached to the satellite for which the Attitude Simulator is computing the attitude,

then, if you specify the keyword RefType with the value Attitude, the sensor's attitude will be computed relative to the attitude state information that is being integrated by the Attitude Simulator.

## Access Constraint Plug-in Points

STK already provides numerous Access constraints that can be used to model visibility between objects. However, new mission proposals often involve new visibility requirements that cannot be readily configured using the existing Access constraints. While this is a prime motivator for incorporating new constraints directly into the STK software, nongeneric constraints are rarely added. Access Constraint Plug-in Scripts provide a method for adding these customized constraints directly into visibility computations.

The access constraint Plug-in Script functionality is designed to be a seamless extension of the basic constraint processing.

When configured for an object, STK processing will utilize the Plug-in Script when computing Access between objects. The Plug-in Script appears as a Min/Max constraint to the rest of the STK processing.

An access constraint Plug-in Script processes inputs (which it has requested) and returns a single result (a double), for each access processing step. The value provides a (continuous) measure of visibility between the FromObject and the ToObject at each requested sample time. This value is utilized by the access algorithms during sampling to predict and detect whether the value crosses the Min or Max value set for the constraint. The classic example of a visibility constraint value is the elevation angle of the ToObject with respect to the FromObject (which of course is already a constraint that STK provides).

> NOTE: The script may request input data from either object involved in an Access computation, using the descriptors that are documented below. Many of the keyword Names use the prefix *from* or *to* to indicate from which object the data should be obtained. In this context, *from* refers to the object on which this plug-in constraint has been applied and *to* refers to the other object: this association of *from* and *to* is independent of the manner in which Access is computed. For example, if this plug-in constraint is set on MyFacility, then the *from* Object is MyFacility even when the Access is computed from MySatellite to MyFacility.

There are three steps to utilizing an access constraint Plug-in Script:

1. The script must be written to conform to a well-defined calling convention.
2. The script source file must be placed in a specific directory path, which is searched at STK application startup.
3. A particular object's constraints must be configured to use the Plug-in Script by setting the Min and/or Max value in the Constraint Property panel.

The first two steps make the constraint available for certain classes of objects.

## Script Source File Location

The script source files must reside in the proper directory search path to be able to utilize the Plug-in Script functionality. They must be located in one of the following directories:

(i)  <STK install folder>/STKData/Scripting/Constraints

(ii)  <STK user area>/Scripting/Constraints

where <STK install folder> refers to the directory path that is the parent of STKData,and <STK user area> refers to the user's configuration directory.

> **WARNING:** Do not place your scripts directly in the Scenario folder. Doing so may cause errors with your script.

## Additional Calling Modes

In addition to the standard *register* and *compute* calling modes for Plug-in Scripts, constraint Plug-in Points utilize two additional modes: *GetConstraintDisplayName* and *GetAccessList*.

   During STK application start-up, the Scripting/Constraints directories are searched for Plug-in Scripts. Each script is called in *GetConstraintDisplayName* and *GetAccessList* calling modes, and then the constraint is registered as an available constraint in the STK application. Registration of constraints happens only once, on application start-up: while editing the script file will affect subsequent calls in the *register* and *compute* calling modes, *GetConstraintDisplayName* and *GetAccessList* are never called after registration, and any edits to these functions during an STK session have no effect.

### Get Constraint Display Name Method

The *GetConstraintDisplayName* mode is used to ask the script for a String containing the name of the constraint (this name is used on the Constrain Property Panel, for save/load, and for STK/Connect). It is not passed a *methodData* parameter. The *methodData* value is undefined. The name may contain underscores but spaces and special characters are not allowed. The alphabetic case of the returned value is utilized.

### Get Access List Method

The *GetAccessList* method is called with a String *methodData* value containing an STK class name (i.e., the class name of a FromObject). The method is expected to return a String value containing a list of STK classes that are allowed to be a ToObject when used with this constraint. The output String is a list of STK classes separated by commas, e.g.,

```
1. Satellite,Facility,GroundVehicle
```

The returned class names are case sensitive and must exactly match valid class names to be implemented. The *GetAccessList* method is called once for each class that is registered in STK, with a different class name as the input each time. This call has been provided so that the script can restrict applicability of its computations to certain FromObject/ToObject class pairs (e.g., if the script's value is computed assuming that both the FromObject and the ToObject are fixed on the Earth, then the script can restrict its applicability to the Facility and Target classes).

## Constraint Property Panel

When the Plug-in Script constraint is configured properly, it will appear in the Constraint properties, under the name given by the *GetConstraintDisplayName* call, for the object classes that it indicated in the *GetAccessList* method call. The Plug-in tab will list all Plug-in constraints available for the Class. Select a constraint in the Plug-in tab and set a Min and/or Max value for the constraint. Click Change and be sure to Click Apply before leaving the panel.

> NOTE: You must click the Change button for edits to take affect—making edits and clicking Apply is insufficient.

## Inputs and Outputs

The Access Constraint Plug-in Point has the following available inputs and outputs (see Table 35).

## Using the Vector Geometry Tool as an Input Source

In addition to those descriptors shown in Table 35, every input that is available from the Vector Geometry Tool is also an available input for a constraint Plug-in Script. To request an input from the Vector Geometry Tool, simply create a descriptor in the same manner as indicated in the documentation for requesting inputs from that tool.

Moreover, when creating Vector Geometry Tool descriptors, one can use the notation "Access/toObj" and "Access/fromObj" as the value for the keywords Source and RefSource. Example:

```
Type = Axes; Name = Body; Source = Access/toObj.
```

> NOTE: When using an Access Plug-in Constraint as a Figure of Merit, the use of Access/toObj and Access/fromObj should be avoided since the Plug-in will not be reinitilialized for each new Coverage Asset before computing to that asset. Instead, the plug-in will be initialized for all assets *before* computing for any, effectively resolving the object names using the last asset being considered.

**TABLE 35**

Access Constraint Plug-in Point: Inputs & Outputs

| Keyword | ArgumentType | Name | RefName |
|---------|--------------|------|---------|
| Value | Output | Status | — |
| Value | Output | Result | — |
| Value | Output | maxRelMotion | — |
| Value | Input | Epoch | — |
| Value | Input | fromObjectPath | — |
| Value | Input | toObjectPath | — |
| Value | Input | fromCbName | — |
| Value | Input | toCbName | — |
| Value | Input | range | — |
| Value | Input | timeDelay | — |
| Value | Input | fromPosition | Fixed Inertial |
| Value | Input | toPosition | Fixed Inertial |
| Value | Input | fromRelPosition | Fixed Inertial |
| Value | Input | toRelPosition | Fixed Inertial |
| Value | Input | fromVelocity | Fixed Inertial |
| Value | Input | toVelocity | Fixed Inertial |
| Value | Input | fromRelVelocity | Fixed Inertial |
| Value | Input | toRelVelocity | Fixed Inertial |
| Value | Input | fromAngularVelocity | Fixed Inertial |
| Value | Input | toAngularVelocity | Fixed Inertial |
| Value | Input | fromQuaternion | Fixed Inertial |
| Value | Input | toQuaternion | Fixed Inertial |
| Value | Input | fromCbAppSunPosition | Fixed Inertial |
| Value | Input | toCbAppSunPosition | Fixed |

## The Status Output

The Status return value supports two types of segments: message and control. The message segment allows the script to send messages into the STK display window, and the control segment allows the script to control its use in the access calculation. Multiple segments are passed by separating the segments with semicolons.

A control segment has the form:

```
CONTROL: <controlType>;
```

where <controlType> can be one of the control type keywords: OK, STOP, or ERROR. Example:

```
CONTROL: Stop;
```

Control of Stop or Error will cause an error message indicating the problem and the Plug-in will be turned off for the remainder of the run. It is not necessary to return OK on every successful iteration step. The Plug-in is assumed to have succeeded when no control segment is returned.

A message segment has the form:

```
MESSAGE: [<messageType>] <message>;
```

where <messageType> can be one of the message type keywords: ALARM, WARNING, or INFO. Alarms cause the message to be displayed in the message window and the window pops to the front. The <message> value can be any character string terminated by a semicolon. Example:

```
MESSAGE: [Alarm] My Plug-in had an unexpected error!;
```

### Information on Constraint Plug-in Point Arguments

The following table gives representation (data type) and unit information for each constraint Plug-in Point argument, together with a brief definition, (see Table 36).

Velocities pulled from the Access plug-in arguments *fromVelocity* and *toVelocity* are given as observed in the requested frame. The velocities pulled from the Vector Geometry Tool are given as expressed in the requested frame. The observed vector has the angular motion of the two frames taken into account, whereas the expressed vector simply has the components of the base vector rotated into the new frame. Take for example a geostationary satellite. The request for the velocity in the CBF frame will return a value near zero from the Access plug-in argument. If the velocity is pulled from the Vector Geometry Tool, it simply rotates the components of the inertial

**TABLE 36**

Access Constraint Plug-in Point Information

| Name | Representation | Units | Definition |
|---|---|---|---|
| Epoch | Double | epoch secs | The STK epoch time for the current access iteration step. |
| fromCbName | String | — | The central body name for the from object. |
| fromAngularVelocity | Double:3 | 1/sec | The angular velocity for the from object's body, with respect to the reference frame specified. |
| fromCbAppSunPosition | Double:3 | M | The apparent sun position with respect to the center of mass of the from object's central body. |
| fromObjectPath | String | — | The from object path defined in STK fashion. Alternatively, the generic keyword "Access/fromObj" can be used. |
| fromPosition | Double:3 | M | The from object's position in Cartesian coordinate system. |
| fromQuaternion | Double:4 | unitless | The attitude defined by a quaternion for the from object. |
| fromRelPosition | Double:3 | M | The to object's relative position in relation to the from object, in the Cartesian coordinate system of the central body of the from object. |
| fromRelVelocity | Double:3 | m/sec | The to object's relative velocity in relation to the from object, in the Cartesian coordinate system of the central body of the from object. |
| fromVelocity | Double:3 | m/sec | The from object's velocity in Cartesian coordinate system (see below). |
| maxRelMotion | Double | radians | The maximum relative angular motion that the object should move between access iteration step samples (in radians). This number affects constraint sampling. This is an advanced feature and in most cases should not be used. |

*(continued)*

**TABLE 36**

Access Constraint Plug-in Point Information (continued)

| Name | Representation | Units | Definition |
|---|---|---|---|
| range | Double | M | The range between the from and to objects. |
| Result | Double | unitless | The result of the constraint script Plug-in calculations, which will be compared to any minimum and/or maximum constraint values in effect. |
| Status | String | — | A status string that is parsed for message or control segments. Each segment is separated by semicolons. |
| timeDelay | Double | secs | The time delay between the from and to objects. |
| toAngularVelocity | Double:3 | 1/sec | The angular velocity for the to object's body, with respect to the reference frame specified. |
| toCbAppSunPosition | Double:3 | M | The apparent sun position with respect to the center of mass of the to object's central body. |
| toCbName | String | — | The central body name for the to object. |
| toObjectPath | String | — | The to object path defined in STK fashion. Alternatively, the generic keyword "Access/toObj" can be used. |
| toPosition | Double:3 | M | The to object's position in the Cartesian coordinate system |
| toQuaternion | Double:4 | unitless | The attitude defined by a quaternion for the to object |
| toRelPosition | Double:3 | M | The from object's relative position in relation to the to object, in the Cartesian coordinate system of the central body of the to object. |
| toRelVelocity | Double:3 | m/sec | The from object's relative velocity in relation to the to object, in the Cartesian coordinate system of the central body of the to object. |
| toVelocity | Double:3 | m/sec | The to object's velocity in Cartesian coordinate system (see below). |

velocity vector into the CBF frame. To get the same number out of the Vector Geometry Tool, one has to explicitly ask for the CBF Velocity vector, which is separately listed in the object's vector list. This is true for all requested frames however.

Expressed Vector Transformation (Vector Geometry Tool):

$$\vec{v}_b = R_{b,a}\vec{v}_a$$

Observed Vector Transformation (Access arguments):

$$\vec{v}_b = R_{b,a}[\vec{v}_a + (\vec{\omega} \times \vec{p}_a)]$$

## Comm Plug-in Points

The following sections provide information on Plug-in points related to the Comm module.

> NOTE: When you create scripts, it is a good idea to save them in the same folder as the STK scenario to which they pertain.

### Source Transmitter Model

#### *Inputs & Outputs*

The Source Transmitter Model Plug-in point has the following available inputs and outputs (see Table 37).

#### *Information on Arguments*

The following table gives representation (data type) and, where applicable, unit information on each of the above inputs and outputs (see Table 38).

The correlation between the integer values of the PolType element and the different polarization types and their required parameters is given in the following table (see Table 39).

### Receiver Model

#### *Inputs & Outputs*

The Receiver Model Plug-in point has the following available inputs and outputs (see Table 40).

**TABLE** 37

Source Transmitter Model Plug-in Point: Inputs & Outputs

| Keyword | ArgumentType | Name |
|---------|--------------|------|
| Value | Input | DateUTC |
| Value | Input | CbName |
| Value | Input | XmtrPosCBF |
| Value | Input | XmtrAttitude |
| Value | Input | RcvrPosCBF |
| Value | Input | RcvrAttitude |
| Value | Output | Frequency |
| Value | Output | Power |
| Value | Output | Gain |
| Value | Output | DataRate |
| Value | Output | Bandwidth |
| Value | Output | Modulation |
| Value | Output | PostTransmitLoss |
| Value | Output | PolType |
| Value | Output | PolRefAxis |
| Value | Output | PolTiltAngle |
| Value | Output | PolAxialRatio |
| Value | Output | UseCDMASpreadGain |
| Value | Output | CDMAGain |

### Information on Arguments

The following table gives representation (data type) and, where applicable, unit information on each of the above inputs and outputs (see Table 41).

## Custom Antenna Gain

### Inputs & Outputs

The Custom Antenna Gain Plug-in point has the following available inputs and outputs (see Table 42).

### Information on Arguments

The following table gives representation (data type) and, where applicable, unit information on each of the above inputs and outputs (see Table 43).

**TABLE 38**

Source Transmitter Model Plug-in Point Information

| Name | Representation | Units | Definition |
|---|---|---|---|
| DateUTC | String | — | The current date and time. |
| CbName | String | — | The scenario central body. |
| EpochSec | Double | sec | The scenario simulation epoch time. |
| XmtrPosCBF | Double:3 | m | The transmitter position in Central Body Fixed coordinates. A vector of doubles, of length 3, corresponding to the X, Y, and Z values. |
| XmtrAttitude | Double:4 | | The attitude quaternion of the transmitter. A vector of doubles, of length 4. |
| RcvrPosCBF | Double:3 | m | The position of the receiver to which the transmitter is linking in the current time step, in Central Body Fixed coordinates. A vector of doubles, of length 3, corresponding to the X, Y, and Z values. |
| RcvrAttitude | Double:4 | | The attitude quaternion of the receiver to which the transmitter is linking in the current time step. A vector of doubles, of length 4. |
| Frequency | Double | Hz | The frequency of the transmitter carrier. |
| Power | Double | dBW | The final output power. |
| Gain | Double | dBi | The gain of the radiating elements of the antenna. |
| DataRate | Double | bits/sec | The information bit rate. |
| Bandwidth | Double | Hz | The bandwidth of the RF spectrum. |
| Modulation | String | | The type of modulation used by the transmitter. This must be one of the STK registered modulation types. However, users can add their own modulation types, and STK will register them (see online Help for the Comm module). |
| PostTransmitLoss | Double | dE | A collection of post-transmit losses, e.g., antenna dish coupling or radome loss. |
| PolType | Integer | — | The type of polarization (see table below). |
| PolRefAxis | Integer | — | The polarization reference axis used to align transmitter polarization to receiver polarization, with 0, 1, and 2 representing the X, Y, and Z axes, respectively. |
| PolTiltAngle | Double | deg | The polarization tilt angle measured from the reference axis. |
| PolAxialRatio | Real | — | The polarization axial ratio. |
| UseCDMASpreadGain | Boolean | — | A flag (0 or 1) that specifies whether or not to use the bandwidth spreading gain. |
| CDMAGain | Double | dB | The CDMA coding gain advantage. |

**TABLE 39**

PolType Values and Polarization Types

| Value | Polarization Type | Required Parameter(s) |
|-------|-------------------|------------------------|
| 0 | None | None |
| 1 | Linear | Reference Axis |
| 2 | Right Hand Circular | None |
| 3 | Left Hand Circular | None |
| 4 | Vertical | Reference Axis, Tilt Angle |
| 5 | Horizontal | Reference Axis, Tilt Angle |
| 6 | Elliptical | Reference Axis, Tilt Angle, Axial Ratio |

**TABLE 40**

Receiver Model Plug-in Point: Inputs & Outputs

| Keyword | ArgumentType | Name |
|---------|--------------|------|
| Value | Input | DateUTC |
| Value | Input | CbName |
| Value | Input | XmtrPosCBF |
| Value | Input | XmtrAttitude |
| Value | Input | RcvrPosCBF |
| Value | Input | RcvrAttitude |
| Value | Output | Frequency |
| Value | Output | Bandwidth |
| Value | Output | Gain |
| Value | Output | PreReceiveLoss |
| Value | Output | PreDemodLoss |
| Value | Output | UseRainModel |
| Value | Output | RainOutagePercent |
| Value | Output | PolType |
| Value | Output | PolRefAxis |
| Value | Output | PolTiltAngle |
| Value | Output | PolAxialRatio |
| Value | Output | ReceiverNoiseFigure |
| Value | Output | CableLoss |
| Value | Output | CableNoiseTemp |
| Value | Output | AntennaNoiseTemp |

**TABLE 41**

Receiver Model Plug-in Point Information

| Name | Representation | Units | Definition |
|---|---|---|---|
| DateUTC | String | — | The current date and time. |
| CbName | String | — | The scenario central body. |
| EpochSec | Double | sec | The scenario simulation epoch time. |
| XmtrPosCBF | Double:3 | m | The position of the transmitter to which the receiver is linking in the current type step, in Central Body Fixed coordinates. A vector of doubles, of length 3, corresponding to the $X$, $Y$, and $Z$ values. |
| XmtrAttitude | Double:4 | — | The attitude quaternion of the transmitter to which the receiver is linking in the current type step. A vector of doubles, of length 4. |
| RcvrPosCBF | Double:3 | m | The receiver position in Central Body Fixed coordinates. A vector of doubles, of length 3, corresponding to the $X$, $Y$, and $Z$ values. |
| RcvrAttitude | Double:4 | — | The attitude quaternion of the receiver. A vector of doubles, of length 4. |
| Frequency | Double | Hz | The frequency of the transmitter carrier. |
| Bandwidth | Double | Hz | The bandwidth of the RF spectrum. |
| Gain | Double | dBi | The gain of the radiating elements of the antenna. |
| PreReceiveLoss | Double | dB | A collection of pre-receive losses. |
| PreDemodLoss | Double | dB | A collection of pre-demodulation losses. |
| UseRainModel | Boolean | — | A flag (0 or 1) that specifies whether a rain model is to be used. |
| RainOutagePercent | Double | — | The percent outage value that can be tolerated by the link. |
| PolType | Integer | — | The type of polarization (see Table 39). |
| PolRefAxis | Integer | — | The polarization reference axis used to align transmitter polarization to receiver polarization, with 0, 1, and 2 representing the $X$, $Y$, and $Z$ axes, respectively. |
| PolTiltAngle | Double | deg | The polarization tilt angle measured from the reference axis. |
| PolAxialRatio | Real | — | The polarization axial ratio. |
| RcvrNoiseFigure | Double | dB | The receiver pre-amplifier noise figure. |
| CableLoss | Double | dB | The cable loss value. |
| CableNoiseTemp | Double | deg K | The temperature of the cable. |
| AntennaNoiseTemp | Double | deg K | The receiver antenna noise temperature. |

**TABLE 42**

Custom Antenna Gain Plug-in Point: Inputs & Outputs

| Keyword | ArgumentType | Name |
|---------|--------------|------|
| Value | Input | EpochSec |
| Value | Input | DateUTC |
| Value | Input | CbName |
| Value | Input | Frequency |
| Value | Input | AntennaPosLLA |
| Value | Input | AzimuthAngle |
| Value | Input | ElevationAngle |
| Value | Output | AntennaGain |
| Value | Output | AntennaMaxGain |
| Value | Output | Beamwidth |
| Value | Output | IntegratedGain |
| Value | Output | DynamicGain |

**TABLE 43**

Custom Antenna Gain Plug-in Point Information

| Name | Representation | Units | Definition |
|------|---------------|-------|------------|
| EpochSec | Double | sec | The scenario simulation epoch time. |
| DateUTC | String | — | The current date and time. |
| CbName | String | — | The scenario central body. |
| Frequency | Double | Hz | The current frequency at which antenna gain is desired. |
| AntennaPosLLA | Double:3 | deg, deg, m | A vector of doubles, of length 3, representing latitude, longitude, and altitude of the antenna above the surface of the Earth. |
| AzimuthAngle | Double | rad | The azimuth angle measured from the antenna boresight in the antenna rectangular coordinate system used by STK. In combination with ElevationAngle, represents the direction of the comm. link where the gain value is required. |
| ElevationAngle | Double | rad | The elevation angle measured from the antenna boresight in the antenna rectangular coordinate system used by STK. In combination with AzimuthAngle, represents the direction of the comm. link where the gain value is required. |
| AntennaGain | Double | dBi | The antenna gain value in the direction given by the azimuth and elevation angles of the antenna boresight. |

*(continued)*

TABLE 43

Custom Antenna Gain Plug-in Point Information (continued)

| Name | Representation | Units | Definition |
|------|----------------|-------|------------|
| AntennaMaxGain | Double | dBi | The maximum gain of the antenna beam. This value may be at the boresight. |
| Beamwidth | Double | rad | The 3dB beamwidth of the antenna gain pattern. |
| IntegratedGain | Double | — | The antenna integrated gain over a noise source. This value is currently estimated by STK using internal numerical methods. Check the online Help for your version to determine whether it allows a user-supplied value. |
| DynamicGain | Boolean | — | A flag (0 or 1, default = 0) to indicate whether the plug-in is time-based or otherwise contains dynamic data, in which case STK must be notified to recalculate the data as necessary. |

## Absorption Loss Model

### *Inputs & Outputs*

The Absorption Loss Model Plug-in point has the following available inputs and outputs (see Table 44).

### *Information on Arguments*

The following table gives representation (data type) and, where applicable, unit information on each of the above inputs and outputs (see Table 45).

TABLE 44

Absorption Loss Model Plug-in Point: Inputs & Outputs

| Keyword | ArgumentType | Name |
|---------|--------------|------|
| Value | Input | EpochSec |
| Value | Input | DateUTC |
| Value | Input | CbName |
| Value | Input | Frequency |
| Value | Input | XmtrPosCBF |
| Value | Input | RcvrPosCBF |
| Value | Input | XmtrPath |
| Value | Input | RcvrPath |
| Value | Output | AbsorpLoss |
| Value | Output | NoiseTemp |

**TABLE 45**

Absorption Loss Model Plug-in Point Information

| Name | Representation | Units | Definition |
|------|----------------|-------|------------|
| EpochSec | Double | sec | The scenario simulation epoch time. |
| DateUTC | String | – | The current date and time. |
| CbName | String | – | The scenario central body. |
| Frequency | Double | Hz | The Comm link frequency at the current time instant, at which the link budget analysis is being carried out. This would also be the Doppler shifted frequency at the receiver, which is using the frequency to compute propagation losses and noise temperatures to compute its g/T. |
| XmtrPosCBF | Double:3 | m | The transmitter position in Central Body Fixed coordinates. A vector of doubles, of length 3, corresponding to the X, Y, and Z values. |
| RcvrPosCBF | Double:3 | m | The receiver position in Central Body Fixed coordinates. A vector of doubles, of length 3, corresponding to the X, Y, and Z values. |
| XmtrPath | String | – | A string representing the complete scenario path of the current transmitter object. |
| RcvrPath | String | – | A string representing the complete scenario path of the current receiver object. |
| AbsorpLoss | Double | dB | The propagation loss term. |
| NoiseTemp | Double | deg K | The noise temperature associated with the propagation loss, used by STK to compute the antenna noise temperature and the receiver g/T values. |

## Rain Loss Model

### Inputs & Outputs

The Rain Loss Model Plug-in point has the following available inputs and outputs (see Table 46).

### Information on Arguments

The following table gives representation (data type) and, where applicable, unit information on each of the above inputs and outputs (see Table 47).

## Antenna Multibeam Selection Strategy

### Inputs & Outputs

The Antenna Multibeam Selection Strategy Plug-in point has the following available inputs and outputs (see Table 48).

**TABLE 46**

Rain Loss Model Plug-in Point: Inputs & Outputs

| Keyword | ArgumentType | Name |
|---------|--------------|------|
| Value | Input | EpochSec |
| Value | Input | DateUTC |
| Value | Input | CbName |
| Value | Input | Frequency |
| Value | Input | ElevAngle |
| Value | Input | OutagePercentage |
| Value | Input | RcvrPosLLA |
| Value | Input | XmtrPosLLA |
| Value | Output | RainLoss |
| Value | Output | RainNoiseTemp |

**TABLE 47**

Rain Loss Model Plug-in Point Information

| Name | Representation | Units | Definition |
|------|----------------|-------|------------|
| EpochSec | Double | Sec | The scenario simulation epoch time. |
| DateUTC | String | — | The current date and time. |
| CbName | String | — | The scenario central body. |
| Frequency | Double | Hz | The Comm link frequency at the current time instant, at which the link budget analysis is being carried out. This would also be the Doppler shifted frequency at the receiver, which is using the frequency to compute propagation losses and noise temperatures to compute its $g/T$. |
| ElevAngle | Double | Deg | The communication link path elevation angle from the receiver to the transmitter. |
| OutagePercentage | Double | — | The percent of the time that the communication link can tolerate outage. |
| RcvrPosLLA | Double:3 | deg, deg, m | A vector of doubles, of length 3, representing latitude, longitude, and altitude of the receiver above the surface of the Earth. |
| XmtrPosLLA | Double:3 | deg, deg, m | A vector of doubles, of length 3, representing latitude, longitude and altitude of the transmitter above the surface of the Earth. |
| RainLoss | Double | dB | The loss value due to rain. |
| RainNoiseTemp | Double | deg K | The noise temperature associated with the rain loss. |

**TABLE 48**

Antenna Multibeam Selection Strategy Plug-in Point:
Inputs & Outputs

| Keyword | ArgumentType | Name |
|---------|--------------|------|
| Value | Input | EpochSec |
| Value | Input | DateUTC |
| Value | Input | CbName |
| Value | Input | AntennaPosLLA |
| Value | Input | BeamIDsArray |
| Value | Input | NumberOfBeams |
| Value | Input | Frequency |
| Value | Input | Power |
| Value | Input | IsActive |
| Value | Output | BeamNumber |

### Information on Arguments

The following table gives representation (data type) and, where applicable, unit information on each of the above inputs and outputs: (see Table 49).

## CommSystem Link Selection Strategy

### Inputs & Outputs

The CommSystem Link Selection Strategy Plug-in point has the following available inputs and outputs (see Table 50).

### Information on Arguments

The following table gives representation (data type) and, where applicable, unit information on each of the above inputs and outputs (see Table 51).

## Satellite Selection Merit Value

Each transmitter-receiver link pair is identified to the script using index values representing positions in the arrays of transmitters and receivers. Information about the attributes of all transmitters and receivers is passed to the script in the form of arrays. The script is called for each transmitter-receiver link pair available at each time instant. The expected output is the rank order of these links on a relative scale. STK then uses the links beginning with the highest ranked.

**TABLE 49**

Antenna Multibeam Selection Strategy Plug-in Point Information

| Name | Representation | Units | Definition |
|------|----------------|-------|------------|
| EpochSec | Double | sec | The scenario simulation epoch time. |
| DateUTC | String | — | The current date and time. |
| CbName | String | — | The scenario central body. |
| AntennaPosLLA | Double:3 | deg, deg, m | A vector of doubles, of length 3, representing latitude, longitude, and altitude of the antenna above the surface of the Earth. |
| BeamIDsArray | Char | — | A two-dimensional array of characters. The array has a fixed width (number of columns) of 64, and the length of the array (number of rows) is equal to the number of beams available in the multibeam antenna. The array is parsed by the script to obtain the beam IDs. Each row of the character array contains the BeamID of one beam. |
| NumberOfBeams | Integer | — | The total number of beams available in the multibeam antenna at each time instant. This also represents the length of arrays that contain the beam characteristic information, e.g., ID, frequency, power, etc. |
| Frequency | Double | Hz | An array of real values that contains the beam frequency information for each beam. The length of the beam array is equal to the NumberOfBeams element. |
| Power | Double | dBW | For multibeam antennas attached to transmitters. These will have an additional array containing the power value for each beam. The receiver antenna beam power array contains all zero values. |
| IsActive | Integer | — | An array of integers indicating whether each beam is active (1) or inactive (0) in the multibeam model. |
| BeamNumber | Integer | — | The serial number of the beam, in the range $1-n$, where $n = $ NumberOfBeams, to be used for link computations. |

**TABLE 50**

CommSystem Link Selection Strategy Plug-in Point:
Inputs & Outputs

| Keyword | ArgumentType | Name |
|---------|--------------|------|
| Value | Input | DateUTC |
| Value | Input | EpochSec |
| Value | Input | CbName |
| Value | Input | CommSysPath |
| Value | Input | FromIndex |
| Value | Input | NumberOfFromObjects |
| Value | Input | FromObjectsIDArray |
| Value | Input | FromObjectIsStatic |
| Value | Input | FromObjectPosCBFArray |
| Value | Input | FromObjectPosLLAArray |
| Value | Input | FromToRelPosArray |
| Value | Input | FromObjectAttitudeArray |
| Value | Input | ToIndex |
| Value | Input | NumberOfToObjects |
| Value | Input | ToObjectsIDArray |
| Value | Input | ToObjectIsStatic |
| Value | Input | ToObjectPosCBFArray |
| Value | Input | ToObjectPosLLAArray |
| Value | Input | ToFromRelPosArray |
| Value | Input | ToObjectAttitudeArray |
| Value | Output | SatSelMeritValue |

## Comm Constraint

The following sections contain information on a plug-in that applies Comm Link Budget criteria to determine whether access exists.

### Inputs & Outputs

The Comm Constraint Plug-in point has the following available inputs and outputs (see Table 52).

### Information on Arguments

The following table gives representation (data type) and, where applicable, unit information on each of the above inputs and outputs (see Table 53).

**TABLE 51**

CommSystem Link Selection Strategy Plug-in Point Information

| Name | Representation | Units | Definition |
|---|---|---|---|
| DateUTC | String | — | The current date and time. |
| EpochSec | Double | sec | The scenario simulation epoch time. |
| CbName | String | — | The scenario central body. |
| CommSysPath | String | — | The full path and name of the CommSystem in the scenario that defines the interference environment and contains the link selection strategy. |
| FromIndex | Integer | — | The index number of the transmitter from the array of transmitters that is currently being used as the link object. Starts at 0. |
| NumberOfFromObjects | Integer | — | The total number of transmitters in the transmitter array being passed to the script. |
| FromObjectsIDArray | Char | — | The IDs of the transmitters in the array containing information about all the transmitters. |
| FromObjectIsStatic | Boolean | — | Flag indicating whether the transmitter object is static, i.e., fixed on the ground in position and attitude. |
| FromObjectPosCBFArray | Double:3 | m | The positions of the transmitters in the CBF Cartesian coordinate frame. Each element of the array gives position for one transmitter. |
| FromObjectPosLLAArray | Double:3 | deg, deg, m | The positions of the transmitters in latitude, longitude, and altitude. Each element of the array gives position for one transmitter. |
| FromToRelPosArray | Double:3 | m | The relative positions of the transmitters with respect to the current receiver object in the Cartesian coordinate frame. |

*(continued)*

**TABLE 51**

CommSystem Link Selection Strategy Plug-in Point Information (continued)

| Name | Representation | Units | Definition |
|---|---|---|---|
| FromObjectAttitudeArray | Double:4 | — | An array of attitude quaternions (vectors of length 4) of the transmitters. |
| ToIndex | Integer | — | The index number of the receiver from the array of receivers that is currently being used as the link object. Starts at 0. |
| NumberOfToObjects | Integer | — | The total number of receivers in the receiver array being passed to the script. |
| ToObjectsIDArray | Char | — | The IDs of the receivers in the array containing information about all the receivers. |
| ToObjectIsStatic | Boolean | — | Flag indicating whether the receiver object is static, i.e., fixed on the ground in position and attitude. |
| ToObjectPosCBFArray | Double:3 | m | The positions of the receivers in the CBF Cartesian coordinate frame. Each element of the array gives position for one receiver. |
| ToObjectPosLLAArray | Double:3 | deg, deg, m | The positions of the receivers in latitude, longitude, and altitude. Each element of the array gives position for one receiver. |
| ToFromRelPosArray | Double:3 | m | The relative positions of the receivers with respect to the current receiver object in the Cartesian coordinate frame. |
| ToObjectAttitudeArray | Double:4 | — | An array of attitude quaternions (vectors of length 4) of the receivers. |
| SatSelMeritValue | TBD | — | The relative merit value (see below) assigned to the link being analyzed. The scale is open-ended. |

**TABLE 52**

Comm Constraint Plug-in Point: Inputs & Outputs

| Keyword | ArgumentType | Name |
|---------|--------------|------|
| Value | Input | Date |
| Value | Input | DateUTC |
| Value | Input | EpochSec |
| Value | Input | CbName |
| Value | Input | ReceiverPath |
| Value | Input | TransmitterPath |
| Value | Input | RcvrPosCBF |
| Value | Input | RcvrAttitude |
| Value | Input | XmtrPosCBF |
| Value | Input | XmtrAttitude |
| Value | Input | ReceivedFrequency |
| Value | Input | DataRate |
| Value | Input | Bandwidth |
| Value | Input | CDMAGainValue |
| Value | Input | ReceiverGain |
| Value | Input | PolEfficiency |
| Value | Input | PolRelativeAngle |
| Value | Input | RIP |
| Value | Input | FluxDensity |
| Value | Input | GOverT |
| Value | Input | CarrierPower |
| Value | Input | BandwidthOverlap |
| Value | Input | Cno |
| Value | Input | CNR |
| Value | Input | EbNo |
| Value | Input | BER |
| Value | Output | Plug-inConstraintValue |

## Troubleshooting

### Syntax Errors

By far, the most common problem that will be encountered with the use of scripts is a script-compilation problem related to a syntax error. While STK will attempt to report errors that it finds, errors in the scripts themselves are much more difficult to diagnose since they arise not in STK per se but in the language's processing of the script. Often, the language detects an error but will not identify the cause of the problem back to STK.

We **strongly recommend** that users check for syntax and compilation errors in the scripts **before** using them with an STK Plug-in Script.

**TABLE 53**

Comm Constraint Plug-in Point Information

| Name | Representation | Units | Definition |
|---|---|---|---|
| DateUTC | String | — | The current date and time. |
| EpochSec | Double | Sec | The scenario simulation epoch time. |
| CbName | String | — | The scenario central body. |
| ReceiverPath | String | — | A string representing the complete scenario path of the current receiver object. |
| TransmitterPath | String | — | A string representing the complete scenario path of the current transmitter object. |
| RcvrPosCBF | Double:3 | M | The receiver position, in Central Body Fixed coordinates. A vector of doubles, of length 3, corresponding to the X, Y, and Z values. |
| RcvrAttitude | Double:4 | — | The attitude quaternion of the receiver. A vector of doubles, of length 4. |
| XmtrPosCBF | Double:3 | M | The transmitter position in Central Body Fixed coordinates. A vector of doubles, of length 3, corresponding to the X, Y, and Z values. |
| XmtrAttitude | Double:4 | — | The attitude quaternion of the transmitter. A vector of doubles, of length 4. |
| ReceivedFrequency | Double | Hz | The frequency of the signal as seen by the receiver. This value includes any Doppler shift. |
| DataRate | Double | Bits/sec | The information bit rate. |
| Bandwidth | Double | Hz | The bandwidth of the RF spectrum. |
| CDMAGainValue | Double | dB | The CDMA coding gain advantage. |
| ReceiverGain | Double | Db | The antenna gain in the direction of the transmitter. |
| PolEfficiency | Double | — | Polarization efficiency is the polarization mismatch loss due to the misalignment of the reference axes of the transmitter and the receiver. It is expressed or a scale of 0–1. |

*(continued)*

**TABLE 53**

Comm Constraint Plug-in Point Information (continued)

| Name | Representation | Units | Definition |
|------|----------------|-------|------------|
| PolRelativeAngle | Double | Rad | The Polarization Relative angle is the relative angle between the transmitter's polarization reference axis and the receiver's polarization reference axis. The angles are computed after taking into consideration the attitudes of the all of the antenna's parent objects, the orientations of the sensors (if any) and the orientation of the antennas with respect to the transmitter or receiver's body axis |
| RIP | Double | dB W/m² | The Received Isotropic Power at the receiver antenna. |
| FluxDensity | Double | dB·W/m² Hz | The Power flux density at the receiver antenna. |
| GOverT | Double | dB/K | The Receiver gain over the equivalent noise temperature. |
| CarrierPower | Double | dBW | The Carrier power at the receiver input. This is the power received at the receiver LNA input. It takes into consideration pre-receiver losses, antenna gain, cable losses, etc. |
| BandwidthOverlap | Double | — | This is a fraction from 0 to 1 representing the amount of overlap between the transmitted signal and the receiver bandwidths. The amount of power received by the receiver is equal to the transmitted EIRP multiplied by the bandwidth overlap and taking into account any propagation losses. |
| Cno | Double | dB | The Carrier-to-Noise density at the receiver input. |
| CNR | Double | dB | The Carrier-to-Noise ratio at the receiver input. |
| EbNo | Double | dB | The Signal-to-Noise ratio at the receiver. |
| BER | Double | — | The Bit Error Rate. |
| Plug-inConstraintValue | Double | — | The computed constraint value, which is compared to the specified Min and Max values to determine whether access exists. |

## Checking MATLAB Scripts

To check a MATLAB.m file, you will need to run the script within MATLAB in two modes: *register* and *compute*. You will need to create the correct input struct for each case.

## Checking VBScript Scripts

To check a VBScript script, simply double-click on it in the Explorer browser: any compilation errors will be reported in a pop-up window. If no errors are detected, then no indication is given, and you can assume that no compilation problems were encountered.

There still may be a run-time problem—the Microsoft script debugger can be used to diagnose these problems.

## Checking Perl Scripts

To check a Perl script, open a Command Prompt and run the Perl script as you would normally do. Compilation errors will be printed in the window.

## Checking for Proper Registration of Inputs and Outputs

Both VBScript and Perl provide a mechanism to print out (or pop up a panel containing) a list of inputs and outputs, along with their representation. This can be used during script development. Once the script has been shown to work correctly, simply comment out the lines that print this information.

## Checking for Valid values

To check for valid values, simply print them out every so often during the *compute* call, either by saving them to a file or popping up a panel containing the values. We **strongly recommen**d that you do not pop up panels for every call to *compute*, as there may be thousands of calls made to *compute* and you will need to dismiss the panel for each call.

In MATLAB, on a PC, you can access variables in another way. You can declare certain variables global in the .m file and declare them global in the workspace (i.e., at the input prompt). Then, after running, you can print these variables to check whether they are correct or not.

For VBScript, it may be easier to output informational data directly to Microsoft Excel than to save the data to a file.

## Using STK/Connect

Users **cannot** issue STK/Connect commands to STK from inside a Plug-in Script. This causes an infinite loop in the software.

**Further MATLAB Issues**

*File Location*

The MATLAB application locates files to execute by searching the hierarchy of paths. The search path is updated automatically every time a scenario is loaded or created. However, paths are only added to the MATLAB search hierarchy if they exist in the file system. For example, if a new scenario is created that needs to use MATLAB Plug-ins, the following steps must be taken:

1. Create the scenario, and configure everything except for Plug-ins.
2. Save and close the scenario.
3. Create the appropriate subdirectories of the scenario directory.
4. Put the.m files in the appropriate subdirectories.
5. Load the scenario and configure the Plug-ins.

Alternatively, the Plug-ins can be placed in the user configuration area before STK is started, in which case these Plug-ins will be immediately available for every scenario.

*Editing MATLAB Files in Use by STK*

Care must be taken when editing MATLAB files while they are being used by STK. For example, Vector Geometry Tool custom axes and vectors must be reloaded in order for the changes to take effect. In general, only STK-related information will be cleared from the MATLAB workspace. This means that if a MATLAB Plug-in file uses other MATLAB files, changes to those files will not take effect until the user manually clears them from the MATLAB workspace. This can be done by executing clear functionName in the MATLAB command window opened by STK.

*MATLAB Plug-in Scripts on UNIX*

On UNIX systems, MATLAB will be started in the background—no GUI or command prompt is available. To debug MATLAB Plug-in Scripts on UNIX, use a file to store informational data.

# Appendix B: Light Time Delay and Apparent Position

Analytical Graphics, Inc.

## CONTENTS

## Introduction

The relative position of an object B with respect to an object A can be computed in a variety of ways, depending on the model of signal transmission between the two objects. When light time delay is not considered, the speed of light is considered infinite and there is no time difference between the transmission event and the reception event. Quantities that ignore light time delay are often termed "true"; e.g., the true relative position of B with respect to A, is computed as:

$$\mathbf{r}(t) = \mathbf{R}_B(t) - \mathbf{R}_A(t) \tag{1}$$

The term "apparent" is used when the relative position vector accounts in some way for light time delay. The apparent position models signal transmission occurring at the finite speed of light so that a signal transmitted at time $t$ is not received until $t+\Delta t$, where $\Delta t$ is the light time delay (a positive number).

Light propagation models have a rich history, but for our purposes we only need be concerned with three kinematics models: (i) Galilean Relativity, (ii) Special Relativity; and (iii) General Relativity.

Galilean Relativity is by far the most widely known model, where space is completely separable from the concept of time. Space is modeled as a Euclidean space with the standard vector operations for a linear space; time is an absolute quantity known to all observers. Special Relativity models light propagation in such a manner that all inertial observers will measure the speed of light (in vacuum) as the same constant value $c$. Space is no longer separable from time; space-time is not a Euclidean space but instead a Minkowski space. Concepts that were once trivial now become more complicated: different inertial observers now disagree on simultaneity of events, on distances between objects, and even on how fast time evolves. However, light still propagates as a straight line in the spatial components. General Relativity goes one step further, removing the special status of inertial observers and introducing mass as generating the curvature of space-time itself. The light path deflects (curves) in the spatial components near massive objects.

Our goal in the modeling of light propagation is simply to include the first-order corrections on Galilean Relativity caused by Special Relativity for signal transmission. Thus, we strive for accuracy to order $\beta$, where $\beta = v/c$, where $v$ is the inertial velocity of a frame being considered. The light path then is a straight line in inertial space where the signal moves at constant speed $c$ (i.e., gravitational deflection is ignored).

---

## Computing Light Time Delay

We will first consider the light time delay for a signal transmitted from an object A to an object B. Later, we will consider the delay for a received signal at A.

### Transmission from A to B

Consider an inertial frame $F$. Let $\mathbf{R}_A$ locate object A in $F$; let $\mathbf{R}_B$ locate object B in $F$. Let the relative position vector $\mathbf{r}$ be defined by

$$\mathbf{r}(t) = \mathbf{R}_B(t + \Delta t) - \mathbf{R}_A(t) \qquad (2)$$

where $t$ is the time of transmission from A and $t+\Delta t$ is the time of reception at B. The light time delay is $\Delta t$ which will depend on $t$ as well. Let $r = ||\mathbf{r}||$ be the range between the objects. Then $\Delta t = r/c$.

One usually knows the locations for the objects A and B and computes the light time delay at time $t$ through iteration. First, a value of $\Delta t$ is guessed (often taken to be 0.0 or the last value computed at a previous time) and $\mathbf{r}(t)$ is computed. A new value for $\Delta t$ is found from $r/c$ and the procedure repeats. The iteration stops whenever the improvement in the estimate to $\Delta t$ is less than the light time delay convergence tolerance. Typically, few iterations are required as the procedure converges very rapidly.

**Reception at A from Signal Sent from B**

In this case, the relative position vector is

$$\mathbf{r}(t) = \mathbf{R}_B(t - \Delta t) - \mathbf{R}_A(t) \tag{3}$$

where the signal is received by A at time $t$. The same procedure is used to find $\Delta t$, using $\mathbf{r}(t)$ above.

Note: The light time delay $\Delta t$ computed for the transmission from A and for reception at A are different, as is the relative position vector $\mathbf{r}$.

**The Inertial Frame**

The choice of the inertial frame is important when computing light time delay, as it will affect the results. This is a consequence of Special Relativity. Let $F$ and $F'$ be two inertial frames with parallel axes. Let $\mathbf{v}$ be the velocity of $F'$ with respect to $F$. In Special Relativity, time is not absolute but is instead associated with a frame: let $t$ denote time in $F$ and $t'$ denote time in $F'$. For simplicity, assume the frames are coincident at $t = 0$. Then the Lorentz transformation relating these two coordinate time values is

$$t' = \gamma\left(t - \boldsymbol{\beta} \cdot \frac{\mathbf{R}}{c}\right) \tag{4}$$

where $\mathbf{R}$ is the position vector of a location in $F$ (measured from the origin of $F$), and

$$\beta = \frac{v}{c}, \beta = ||\boldsymbol{\beta}||, \delta = \sqrt{1 - \beta^2}, \text{and } \gamma = \frac{1}{\delta} \tag{5}$$

The value $t'$ is the value of time in $F'$ for an event at time $t$ at position $\mathbf{R}$ in $F$. Note that $t'$ depends on both $t$ and $\mathbf{R}$.

Consider the case of transmission from object A, located at the origin of $F$ at time $t = 0$, to object B that receives the signal at time $t = \Delta t$. The value of $t'$ at transmission is computed to be 0 (since both $t$ and $\mathbf{R}$ are zero then). In $F$, the light time delay $\Delta t$ is computed by solving

$$\|\mathbf{R}_B(\Delta t)\| = c\Delta t \tag{6}$$

for $\Delta t$. Using the Lorentz transformation, the value of $t'$ at reception is

$$t' = \gamma\left(\Delta t - \boldsymbol{\beta} \cdot \frac{\mathbf{R}_B}{c}\right) = \gamma\Delta t\left(1 - \hat{\mathbf{e}}_{R_B} \cdot \boldsymbol{\beta}\right), \text{ where } \hat{\mathbf{e}}_{R_B} = \frac{\mathbf{R}_B}{c\Delta t} \tag{7}$$

In $F'$, the light time delay is

$$\Delta t' = \gamma\Delta t\left(1 - \hat{\mathbf{e}}_{R_B} \cdot \boldsymbol{\beta}\right) \tag{8}$$

To order $\beta$, $\gamma$ is 1.0, so the difference $\delta t$ in the computed light time delays between the two frames is

$$\delta t = \Delta t - \Delta t' = \Delta t\left(\hat{\mathbf{e}}_{R_B} \cdot \boldsymbol{\beta}\right) \tag{9}$$

The case of reception at A at time $t = 0$ is analogous, producing the same result.

Note: The choice of inertial frame affects the computation of both the light time delay $\Delta t$ and the apparent relative position vector $\mathbf{r}$.

### Choosing an Inertial Frame

Most space applications involve objects located near one central body. It is natural to associate a central body with an object. The natural inertial frame to use for modeling spacecraft motion near a central body is the inertial frame of the central body. (We use the term CBI for Central Body Inertial.) The CBI frame is a natural choice for the inertial frame for computing light time delay.

For objects that are far from their central body, however, the more appropriate inertial frame to model motion is a frame with origin at the solar system barycenter. This frame is used to model the motion of the central bodies themselves. This provides another choice for the inertial frame.

We take the view that CBI is the preferable frame for computing light time delay, but we want to ensure that its use appropriately models the physics of the situation at hand. Thus, at the start of the light time delay computation we compute the difference $\delta t$ between the use of the CBI and solar system barycenter frames. If this difference is less than the light time delay convergence tolerance, then either frame may be used to obtain the same level of accuracy—we choose the CBI frame because it is less expensive computationally.

If the difference is more than the tolerance, we use the solar system barycenter frame knowing that it is a better model of an inertial frame in general.

## Earth Operations

For the light time delay convergence tolerance of $50^{-5}$ seconds (i.e., 50 microseconds), objects located from near the Earth's surface to just outside the geosynchronous belt will use Earth's inertial frame for performing light time delay computations. Farther out than this, the solar system barycenter frame will be used. In particular, computations involving objects at the Earth–Moon distance will use the solar system barycenter frame for computation of light time delay.

## Signal Path

With $\Delta t$ and $\mathbf{r}$ determined from the light time delay computation performed in the inertial frame $F$, it is now possible to model the actual signal transmission (i.e., the path of the signal through $F$). The signal path is given by

$$\text{Transmit from A at } t: \ \mathbf{s}(t+\tau)=\mathbf{R}_A(t)+c\tau\hat{\mathbf{e}}_r \tag{10}$$

$$\text{Receive at A at } t: \ \mathbf{s}(t+\tau-\Delta t) = \mathbf{R}_A(t)+c(\Delta t-\tau)\hat{\mathbf{e}}_r \tag{11}$$

where $0 < \tau < \Delta t$ and $\tau = 0$ locates the transmission event and $\tau = \Delta t$ locates the reception event. The apparent direction is given by

$$\hat{\mathbf{e}}_r = \frac{\mathbf{r}}{r}, r = \|\mathbf{r}\|, \mathbf{r} = \mathbf{R}_B(t+\sigma\Delta t)-\mathbf{R}_A(t) \tag{12}$$

where $\Delta t$ is the light time delay computed in $F$ and $\sigma = 1$ in the case of transmission and $\sigma = -1$ for reception.

---

## Aberration

Aberration is the change in the perceived direction of motion caused by the observer's own motion. The classic example of aberration involves two men out in the rain. One man is stationary and perceives the velocity of the rain as straight down from overhead at velocity $\mathbf{u}$. The other man is walking along the ground at velocity $\mathbf{v}$. In the moving man's frame, the velocity of the rain is $\mathbf{u} - \mathbf{v}$. (This is the value as computed using Galilean Relativity; the value according to Special Relativity is more complicated but the conclusions are the same). This relative velocity makes an angle $\varphi$ with the vertical where

$$\varphi = \tan^{-1}\left(\frac{v}{u}\right) \tag{13}$$

The faster the man walks, the larger his perceived deflection of the rain from the vertical.

In technical sources, aberration is usually discussed in the context of either stellar or planetary aberration. Stellar aberration was first considered when looking at stars through optical telescopes—it is the perceived change in direction of light. Planetary aberration usually refers to two effects combined, light time delay and the perceived change in the direction of light. In both cases, the observer's velocity relative to the frame in which the light path was computed results in aberration.

## Stellar Aberration

Typically, starlight is modeled as saturating the solar system with light. The light is considered to move in a straight line through the solar system. The actual transmission time at the star is unmodeled (being more uncertain than the direction to the star itself) so light time delay is not considered. However, aberration caused by an observer's motion in the solar system as the observer receives the light can be computed, and is referred to as stellar aberration. Let the direction to a star from an observer (accounting for proper motion of the star and parallax) be $\hat{\mathbf{e}}_r$. Then the apparent direction of the star, accounting for stellar aberration, is:

$$\hat{\mathbf{P}} = \frac{\hat{\mathbf{e}}_r + \beta}{\left\|\hat{\mathbf{e}}_r + \beta\right\|}, \text{ where } \beta = \frac{\mathbf{v}}{c} \tag{14}$$

and $\mathbf{v}$ is the velocity of the observer with respect to the solar system barycenter frame. The formula above is the Galilean formula, Equation (3.252-1) [as given in the *Explanatory Supplement to the Astronomical Almanac*]; the Special Relativity formula is given by (3.252-3) [from the same reference], and is simply a use of the Lorentz transformation for velocities. The formula above is accurate to order $\beta$.

Note: The stellar aberration formula above models the observer receiving a signal, not transmitting a signal.

## Annual and Diurnal Aberration

While the concept of aberration is simple, its computation can be complicated, depending on which factors are considered for determining the observer's velocity $\mathbf{v}$ with respect to $F$. Astronomers have compartmentalized different aspects of the computation, coining terms for each aspect's contribution. The term "annual aberration" is meant to identify the contribution of the observer's central body velocity in the solar system:

$$\mathbf{v} = \mathbf{v}_{cb} + \mathbf{v}_{A/cb} \tag{15}$$

where $\mathbf{v}_{cb}$ is the velocity of the central body with respect to the solar system barycenter frame and $\mathbf{v}_{A/cb}$ is the velocity of the observer A with respect to the central body. When $\mathbf{v}_{cb}$ is used to compute aberration, rather than $\mathbf{v}$, then only the effects of annual aberration have been considered in the proper apparent relative position.

We term $\mathbf{v}_{A/cb}$ the diurnal aberration (the contribution to $\mathbf{v}$ apart from the central body motion). In some technical sources, the term "diurnal aberration" is reserved for the contribution to $\mathbf{v}_{A/cb}$ made by the rotation of the central body itself, and other terms are used to describe the other contributions to the overall value of $\mathbf{v}_{A/cb}$.

## Planetary Aberration

Usually, planetary aberration refers to two effects combined: light time delay and the stellar aberration (i.e., the change in the perceived direction of motion caused by an observer's motion). To order $\beta$, the results can be computed correctly using the simpler Galilean formulas.

We have previously discussed light time delay and determined a method for computing the light time delay $\Delta t$, the apparent relative position vector $\mathbf{r}$, and the signal path $s$ by identifying an inertial frame $F$ to perform the computations. We now consider the effect of aberration.

Consider another inertial frame $F'$ coincident with the observer A at the event time $t$, whose constant velocity $\mathbf{v}$ is the value of the observer's velocity at time $t$. Because the observer's velocity is not (usually) constant in time, we'll associate a new inertial frame $F'$ for each time $t$, calling the collection of inertial frames the comoving inertial frames at A. The apparent position of B with respect to A as perceived by an observer at A at time $t$ but moving with $F'$ (computed by modeling the signal motion in $F$ and then transforming this motion to $F'$) is

$$\mathbf{r}_p = \mathbf{r} - \sigma \Delta t \mathbf{v} = c \Delta t \left( \hat{\mathbf{e}}_r - \boldsymbol{\beta} \right) = \mathbf{R}_B (t + \sigma \Delta t) - \mathbf{R}_A (t) - \sigma \Delta t \dot{\mathbf{R}}_A (t) \tag{16}$$

where $\sigma = 1$ when modeling a signal transmitted from A, and $\sigma = -1$ when modeling a signal received at A. Again, $\mathbf{r}$ is the apparent relative position of

B with respect to A, so that the light path range $r$ is $c\Delta t$ and $\beta = \mathbf{v}/c$. This formula generalizes Equation (3.255-2) [as given in the *Explanatory Supplement to the Astronomical Almanac*] to cases of transmission and reception. The vector $\mathbf{r}_p$ is the proper apparent relative position of B with respect to A, where the term "proper" indicates that this quantity is computed as perceived by A (really, by an observer at A moving in a comoving inertial frame).

When the light time delay $\Delta t$ is small, it is possible to construct alternate representations of planetary aberration that approximate the exact expression (16). Expanding $\mathbf{R}_B$ in a Taylor series in time, we find:

$$\mathbf{R}_B(t+\sigma\Delta t) = \mathbf{R}_B(t)+\sigma\Delta t\dot{\mathbf{R}}_B(t)+\cdots \tag{17}$$

Using (17) in (16), we find:

$$\mathbf{r}_p \doteq \mathbf{R}_B(t)-\mathbf{R}_A(t)+\sigma\Delta t\left\{\dot{\mathbf{R}}_B(t)-\dot{\mathbf{R}}_A(t)\right\} \tag{18}$$

which is Equation (3.255-4) [as given in the *Explanatory Supplement to the Astronomical Almanac*] generalized to both transmit and receive cases. Similarly, expanding $\mathbf{R}_A$ in a Taylor series in time, we find:

$$\mathbf{R}_A(t) = \mathbf{R}_A(t+\sigma\Delta t)-\sigma\Delta t\dot{\mathbf{R}}_A(t+\sigma\Delta t)+\cdots \tag{19}$$

Using (19) in (16), we find:

$$\mathbf{r}_p \doteq \mathbf{R}_B(t+\sigma\Delta t)-\mathbf{R}_A(t+\sigma\Delta t)+\sigma\Delta t\left\{\dot{\mathbf{R}}_A(t+\sigma\Delta t)-\dot{\mathbf{R}}_A(t)\right\} \tag{20}$$

that can be simplified to

$$\mathbf{r}_p \doteq \mathbf{R}_B(t+\sigma\Delta t)-\mathbf{R}_A(t+\sigma\Delta t) \tag{21}$$

when the last expression in (20) is small. (This will be small for small $\Delta t$ and small acceleration of A.) This generalizes Equation (3.255-3) [as given in the *Explanatory Supplement to the Astronomical Almanac*] for transmit and receive cases.

The proper apparent direction is computed to be

$$\hat{p} = \frac{\hat{e}-\sigma\beta}{\left\|\hat{e}_r-\sigma\beta\right\|} \tag{22}$$

that of course agrees with the value computed for stellar aberration in the case of reception. The proper apparent direction depends on $\Delta t$ only through $r$. Also note that the proper apparent range $r_p$ is not the same as the light path range $r$, nor is the proper apparent range the same in the case of transmission and reception. This is consistent with Special Relativity as distances in $F$ and $F'$ differ.

## Optical Measurements

Optical observations of satellite position are made by measuring the apparent satellite location against known stars in the telescope field of view. Observations collected in this manner can be used in determining the orbit of the satellite. These observations are a function of the corrections which have been applied to the star positions. Typically these corrections include such effects as the proper motion and parallax of the stars. Star coordinate corrections may optionally include annual and diurnal aberration due to the motion of the observer. Effects not accounted for in the computation of the star coordinates must be accounted for separately in observation processing. For example, omission of diurnal aberration from the star positions requires a diurnal aberration correction in orbit determination. Regardless of the corrections made to the star catalog, orbit determination must also account for the motion of the satellite during the time it takes for light to travel from the satellite to the observer.

# Appendix C: Flow Diagram for the Transmitter Modulation Settings

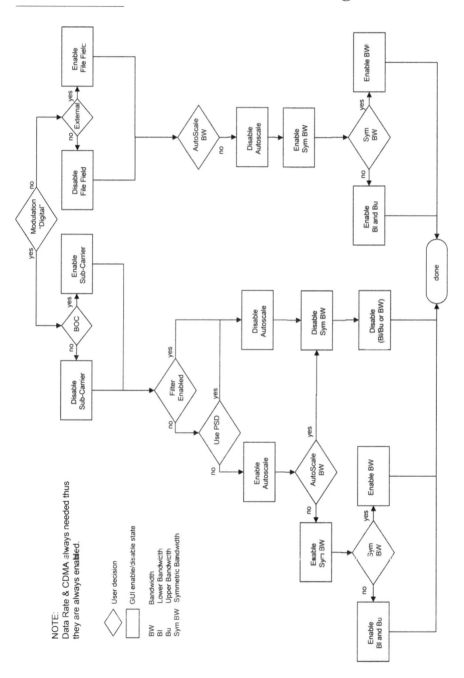

NOTE:
Data Rate & CDMA always needed thus they are always enabled.

◇ User decision

☐ GUI enable/disable state

BW      Bandwidth
Bl      Lower Bandwidth
Bu      Upper Bandwidth
Sym BW  Symmetric Bandwidth

# Index

Printed and bound by CPI Group (UK) Ltd, Croydon, CR0 4YY

01/11/2024

01782625-0005